W9-CAL-854

A VHDL Primer

Third Edition

J. BHASKER

Bell Laboratories, Lucent Technologies
Allentown, PA

Published by Prentice Hall PTR
Upper Saddle River, New Jersey 07458
http://www.phptr.com

Editorial/Production Supervision: *Kathleen M. Caren*
Cover Design Director: *Jerry Votta*
Cover Designer: *Alamini Design*
Manufacturing Manager: *Alan Fischer*
Marketing Manager: *Kaylie Smith*
Acquisitions Editor: *Bernard Goodwin*
Editorial Assistant: *Diane Spina*

Published by Prentice Hall PTR
Prentice-Hall, Inc., A Simon & Schuster Company
Upper Saddle River, New Jersey 07458

Prentice Hall books are widely used by corporations and government agencies for training, marketing, and resale.
The publisher offers discounts on this book when ordered in bulk quantities.
For more information, contact: Corporate Sales Department, Phone: 800-382-3419;
FAX: 201-236-7141; email: corpsales@prenhall.com
Or write: Corp. Sales Dept., Prentice Hall PTR, 1 Lake Street, Upper Saddle River, NJ 07458

Printed in the United States of America
10 9 8 7 6 5 4 3 2

ISBN 0-13-096575-8

Prentice-Hall International (UK) Limited, *London*
Prentice-Hall of Australia Pty. Limited, *Sydney*
Prentice-Hall Canada Inc., *Toronto*
Prentice-Hall Hispanoamericana, S.A., *Mexico*
Prentice-Hall of India Private Limited, *New Delhi*
Prentice-Hall of Japan, Inc., *Tokyo*
Simon & Schuster Asia Pte. Ltd., *Singapore*
Editora Prentice-Hall do Brasil, Ltda., *Rio de Janeiro*

Dedicated to

my parents

Nagamma and Appiah Jayaram

Test Bench
&
complete beh arch
for a state machine

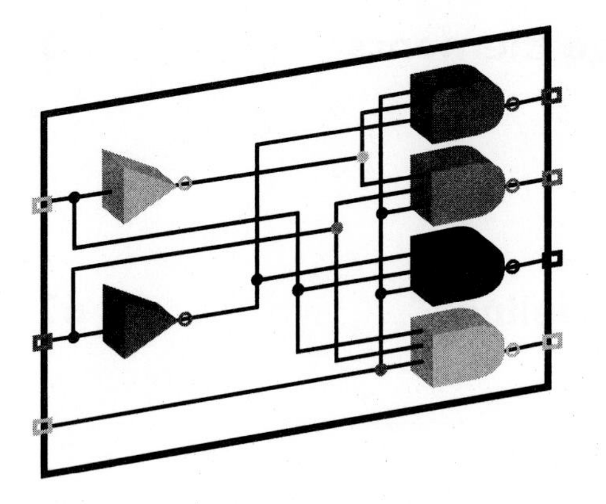

Table of Contents

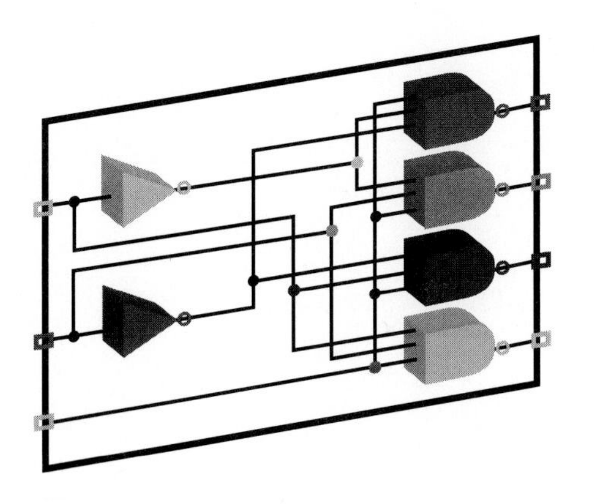

Preface

The IEEE standard STD_LOGIC_1164 package has become very popular and is in widespread use since the last edition came out. This edition of the book replaces the ATT_MVL package with the STD_LOGIC_1164 package. Appendix E provides a listing of the IEEE Std 1164-1993 package.

More examples have been added explaining additional useful features of the language. New sections on text I/O capabilities and test benches have been added. In addition, default parameters in subprograms and binding instances in generate statements are discussed. Also an utility package, UTILS_PKG, is listed in Appendix F. This package is used in some examples of the book where functions were required that were not provided by the STD_LOGIC_1164 package.

The book's format also has been modified to make it more user-friendly and readable.

J. Bhasker
September, 1998

Preface

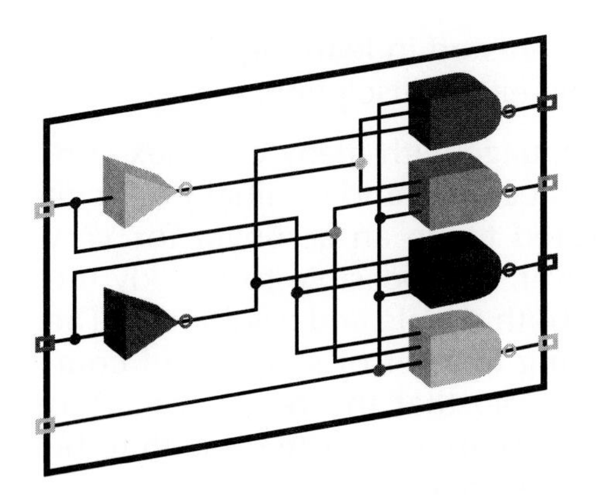

Preface to First Edition

VHDL is a hardware description language that can be used to model a digital system. It contains elements that can be used to describe the behavior or structure of the digital system, with the provision for specifying its timing explicitly. The language provides support for modeling the system hierarchically and also supports top-down and bottom-up design methodologies. The system and its subsystems can be described at any level of abstraction ranging from the architecture level to the gate level. Precise simulation semantics are associated with all the language constructs, and therefore, models written in this language can be verified using a VHDL simulator.

The aim of this book is to introduce the VHDL language to the reader at the beginner's level. No prior knowledge of the language is required. The reader is, however, assumed to have some knowledge of a high-level programming language, like C or Pascal, and a basic understanding of hardware design. This text is intended for both software and hardware designers interested in learning VHDL, with no specific emphasis being placed on either discipline.

VHDL is a large and verbose language with many complex constructs that have complex semantic meanings and is initially difficult to understand (VHDL is often quoted to be an acronym for Very Hard Description Language). However, it is possible to quickly understand a subset of VHDL which is both simple and easy to use. The emphasis of this text is on presenting this set of simple and commonly used features of the language so that the reader can start writing models in VHDL. These features are powerful enough to be able to model designs with high degrees of complexity.

This book is not intended to replace the IEEE Standard VHDL Language Reference Manual, the official language guide, but to complement it by explaining the complex constructs of the language using an example-based approach. Emphasis is placed on providing illustrative examples that explain the different formulations of the language constructs and their semantics. The complete syntax of language constructs is often not described; instead, the most common usage of these constructs is presented. Syntax for constructs is written in a self-explanatory fashion rather than through the use of formal terminology (the Backus-Naur Form), which is used in the Language Reference Manual. This text does not cover the entire language but concentrates on the most useful aspects.

Book Organization

Chapter 1 gives a brief history of the development of the VHDL language and presents its major capabilities. Chapter 2 provides a quick tutorial to demonstrate the primary modeling features. The following chapters expand on the concepts presented in this tutorial. Chapter 3 describes the basic elements of the language such as types, subtypes, objects, and literals. It also presents the predefined operators of the language.

Chapter 4 presents the behavior style of modeling. It introduces the different sequential statements that are available and explains how they may be used to model the sequential behavior of a design. This modeling style is very similar in semantics to that of any high-level programming language. Chapter 5 describes the dataflow style of modeling. It describes concurrent signal assignment statements and block statements, and provides examples to show how the concurrent behavior of a design may be modeled using these statements.

Chapter 6 presents the structural style of modeling. In this modeling style, a design is expressed as a set of interconnected components, possibly in a hierarchy. Component instantiation statements are explained in detail in this chapter. Chapter 7 explains the notion of an entity-architecture pair and describes how component instances can be bound to designs residing in different libraries. This chapter also explains how to pass static information into a design using generics.

Chapter 8 describes subprograms. A subprogram is a function or a procedure. The powerful concept of subprogram and operator overloading is also introduced. Chapter 9 describes packages and the design library environment as defined by the language. It also explains how items stored in one library may be accessed by a design residing in another library. Advanced features of the language such as entity statements, aliases, guarded signals, and attributes are described in Chapter 10.

Chapter 11 describes a methodology for simulating VHDL models and describes techniques for writing test benches. Examples for generating various types of clocks and waveforms and their application to the design under test are presented. Chapter 12 contains a comprehensive set of hardware modeling examples. These include, among others, examples of modeling combinational logic, synchronous logic, and finite-state machines.

In all the VHDL descriptions that appear in this book, reserved words are in **boldface**. A complete list of reserved words also appears in Appendix A. Most of the language constructs are explained using easy-to-understand words, rather than through the use of formal terminology adopted in the Language Reference Manual. Also, some constructs are described only in part to explain specific features. The complete language grammar is provided in Appendix B.

Appendix C contains a complete description of a package that is referred to in Chapters 11 and 12.

In all the language constructs that appear in this book, names in *italics* indicate information to be supplied by the model writer. For example,

entity *entity-name* **is**
    [**port** (*list-of-interface-ports*)] . . .

Entity, **is**, and **port** are reserved words, while *entity-name* and *list-of-interface-ports* represent information to be provided by the model writer. The square brackets, [. . .], indicate optional items. Occasionally, ellipses (. . .) are used in VHDL source to indicate code that is not relevant to that discussion. All examples that are described in this book have been validated using a native VHDL system.

Throughout this text, we shall refer to the circuit, system, design, or whatever it is that we are trying to model as the *entity*.

Book Usage

The first-time reader of this book is strongly urged to read the tutorial presented in Chapter 2. The remaining chapters are organized such that they use information from the previous chapters. However, if the tutorial is well understood, it is possible to go directly to the chapter of interest. A complete chapter has been devoted to examples on hardware modeling; this can be used as a future reference. A list of suggested readings and books on the language and the complete language syntax is provided at the end of the book. For further clarifications on the syntax and semantics of the language constructs, the reader can refer to the IEEE Standard VHDL Language Reference Manual (IEEE Std 1076-1987), published by the IEEE.

Acknowledgments

I gratefully acknowledge the efforts of several of my colleagues in making this book possible. I am especially thankful to Dinesh Bettadapur, Ray Voith, Joel Schoen, Sindhu Xirasagar, Gary Imken, Paul Harper, Oz Levia, Jeff Jones, and Guru Rao. Their constructive criticism and timely review on earlier versions of the text have resulted in several improvements in the book. A special thanks to Gary Imken for being patient enough in answering a number of questions on the idiosyncrasies of VHDL.

I would also like to thank my wife, Geetha, for reviewing the first version of this text and for providing motivational guidance during the entire preparation of this book.

J. Bhasker
October, 1991

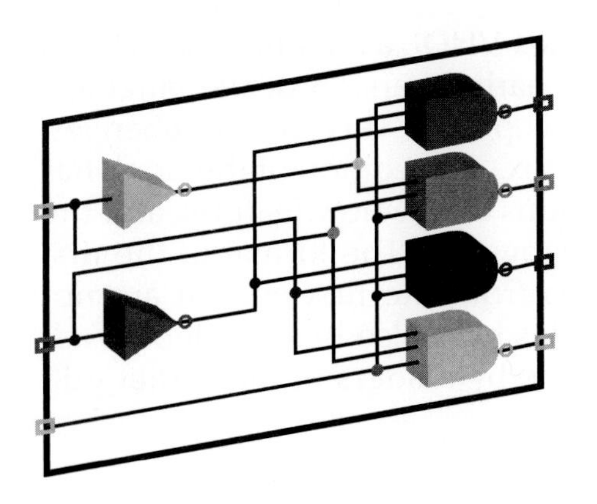

Preface to Second Edition

Synthesis seems to be the driving force in making VHDL a popular hardware description language. As the number of synthesis users has grown, so has the popularity and usage of VHDL. Today, VHDL is an IEEE standard as well as an ANSI standard for describing digital designs.

According to IEEE rules, the language must be reballotted every five years to continue to exist as a standard. VHDL was first standardized in 1987, and the standard was called IEEE Std 1076-1987. The first edition of this book was based on that version of the language. In 1992, the language was reballotted, and after much deliberation, in 1993, a new standard called IEEE Std 1076-1993 was developed. In this version, a number of new features have been added, syntax for certain constructs has been made more consistent, and many ambiguities in the earlier version have been resolved. An appendix in this book summarizes these changes.

This edition of the book describes VHDL as defined in IEEE Std 1076-1993. Models described in the earlier version of the language are strictly upward-compatible, except for very minor changes; these changes are listed in an appendix. This edition of the book has been expanded to include descriptions of the new features and the syntax improvements made to the language. The format of the first edition has been retained. Changes from the first edition are not identified separately. Several examples have also been added. In addition, a number of suggestions made by readers of the earlier edition have been incorporated.

Acknowledgements

I am greatly indebted to Paul Menchini for reviewing this revised edition of the book and providing valuable comments and thoughtful criticism, which have resulted in an improved edition of the book.

J. Bhasker
April, 1994

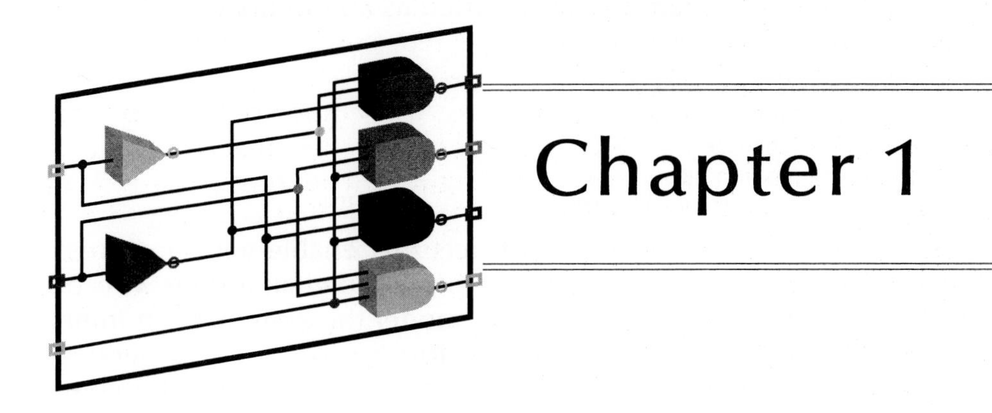

Chapter 1

Introduction

This chapter provides a brief history of the development of VHDL and describes the major capabilities that differentiate it from other hardware description languages. The chapter also explains the concept of an entity.

1.1 What Is VHDL?

VHDL is an acronym for VHSIC Hardware Description Language (VHSIC is an acronym for Very High Speed Integrated Circuits). It is a hardware description language that can be used to model a digital system at many levels of abstraction, ranging from the algorithmic level to the gate level. The complexity of the digital system being modeled could vary from that of a simple gate to a complete digital electronic system, or anything in between. The digital system can also be described hierarchically. Timing can also be explicitly modeled in the same description.

The VHDL language can be regarded as an integrated amalgamation of the following languages:

sequential language +
concurrent language +
net-list language +
timing specifications +
waveform generation language =>VHDL

Therefore, the language has constructs that enable you to express the concurrent or sequential behavior of a digital system with or without timing. It also allows you to model the system as an interconnection of components. Test waveforms can also be generated using the same constructs. All the above constructs may be combined to provide a comprehensive description of the system in a single model.

The language not only defines the syntax but also defines very clear simulation semantics for each language construct. Therefore, models written in this language can be verified using a VHDL simulator. It is a strongly typed language and is often verbose to write. It inherits many of its features, especially the sequential language part, from the Ada programming language. Because VHDL provides an extensive range of modeling capabilities, it is often difficult to understand. Fortunately, it is possible to quickly assimilate a core subset of the language that is both easy and simple to understand without learning the more complex features. This subset is usually sufficient to model most applications. The complete language, however, has sufficient power to capture the descriptions of the most complex chips to a complete electronic system.

1.2 History

The requirements for the language were first generated in 1981 under the VHSIC program. In this program, a number of U.S. companies were involved in designing VHSIC chips for the Department of Defense (DoD). At that time, most of the companies were using different hardware description languages to describe and develop their integrated circuits. As a result, different vendors could not effectively exchange designs with one another. Also, different vendors provided DoD with descriptions of their chips in different hardware description languages. Reprocurement and reuse was also a big issue. Thus, a need for a standardized hardware description language for the design, documentation, and verification of digital systems was generated.

A team of three companies, IBM, Texas Instruments, and Intermetrics, were first awarded the contract by the DoD to develop a version of the language in 1983. Version 7.2 of VHDL was developed and released to the public in 1985. There was strong industry participation throughout the VHDL language development process, especially from the companies that were developing VHSIC chips. After the release of version 7.2, there was an increasing need to make the language an industry-wide standard. Consequently, the language was transferred to the IEEE for standardization in 1986. After a substantial enhancement to the language, made by a team of industry, university, and DoD representatives, the language was standardized by the IEEE in December 1987; this version of the language is known as the IEEE Std 1076-1987. The official language description appears in the IEEE Standard VHDL Language Reference Manual, available from IEEE. The language has also been recognized as an American National Standards Institute (ANSI) standard.

According to IEEE rules, an IEEE standard has to be reballoted every five years so that it may remain a standard. Consequently, the language was upgraded with new features, the syntax of many constructs was made more uniform, and many ambiguities present in the 1987 version of the language were resolved. This new version of the language is known as the IEEE Std 1076-1993. *The language described in this book is based on this standard.* A summary of the major changes made to the 1987 version of the language is presented in Appendix D.

The Department of Defense, since September 1988, requires all its digital Application-Specific Integrated Circuit (ASIC) suppliers to deliver VHDL descriptions of the ASICs and their subcomponents, at both the behavioral and structural levels. Test benches that are used to validate the ASIC chip at all levels in its hierarchy must also be delivered in VHDL. This set of government requirements is described in Military Standard 454.

Since 1987, there has also been a great need for a standard package to aid in model interoperability. This was because different CAE (computer-aided engineering) vendors supported different packages on their systems, causing a major model interoperability problem. Some of the logic values used were 46-value logic, 7-value logic, 4-value logic, and so on. A committee was set up to standardize such a package. The outcome of this committee was the development of a 9-value logic package. This package, called STD_LOGIC_1164, was then balloted and approved to become an IEEE standard, labeled IEEE Std 1164-1993. Some examples in this book use this package.

1.3 Capabilities

The following are the major capabilities that the language provides along with the features that differentiate it from other hardware description languages.

- The language can be *used as an exchange medium between chip vendors and CAD tool users*. Different chip vendors can provide VHDL descriptions of their components to system designers. CAD tool users can use it to capture the behavior of the design at a high level of abstraction for functional simulation.
- The language can also be *used as a communication medium between different CAD and CAE tools*. For example, a schematic capture program may be used to generate a VHDL description for the design, which can be used as an input to a simulation program.
- The language supports *hierarchy*; that is, a digital system can be modeled as a set of interconnected components; each component, in turn, can be modeled as a set of interconnected subcomponents.
- The language supports *flexible design methodologies*: top-down, bottom-up, or mixed.
- The language is *not technology-specific*, but is capable of supporting technology-specific features. It can also support various hardware technologies; for example, you

may define new logic types and new components; you may also specify technology-specific attributes. By being technology-independent, the same model can be synthesized into different vendor libraries.

- It supports both *synchronous and asynchronous timing* models.
- Various digital modeling techniques, such as finite-state machine descriptions, algorithmic descriptions, and Boolean equations, can be modeled using the language.
- The language is publicly available, human-readable, machine-readable, and, above all, it is *not proprietary*.
- It is an *IEEE and ANSI standard*; therefore, models described using this language are portable. The government also has a strong interest in maintaining this as a standard so that reprocurement and second-sourcing may become easier.
- The language supports *three basic different description styles*: structural, dataflow, and behavioral. A design may also be expressed in any combination of these three descriptive styles.
- It supports a wide *range of abstraction levels* ranging from abstract behavioral descriptions to very precise gate-level descriptions. It does not, however, support modeling at or below the transistor level. It allows a design to be captured at a mixed level using a single coherent language.
- *Arbitrarily large designs can be modeled* using the language, and there are no limitations imposed by the language on the size of a design.
- The language has elements that make *large-scale design modeling easier*; for example, components, functions, procedures, and packages.
- *Test benches can be written* using the same language to test other VHDL models.
- Nominal propagation delays, min-max delays, setup and hold timing, timing constraints, and spike detection can all be described very naturally in this language.

- The use of generics and attributes in the models *facilitate back-annotation* of static information such as timing or placement information.
- Generics and attributes are also useful in describing *parameterized designs.*
- A model can not only describe the functionality of a design, but can also contain *information about the design itself* in terms of user-defined attributes, such as total area and speed.
- A common language can be used to describe library components from different vendors. Tools that understand VHDL models will have no difficulty in reading models from a variety of vendors since the language is a standard.
- Models written in this language can be verified by simulation since *precise simulation semantics are defined* for each language construct.
- Behavioral models that conform to a certain synthesis description style are *capable of being synthesized* to gate-level descriptions.
- The *capability of defining new data types* provides the power to describe and simulate a new design technique at a very high level of abstraction without any concern about the implementation details.

1.4 Hardware Abstraction

VHDL is used to describe a model for a digital hardware device. This model specifies the external view of the device and one or more internal views. The internal view of the device specifies the functionality or structure, while the external view specifies the interface of the device through which it communicates with the other models in its environment. Figure 1-1 shows the hardware device and the corresponding software model.

The device-to-device model mapping is strictly one-to-many. That is, a hardware device may have many device models. For example, a device modeled at a high level of abstraction may not have a clock as one of its inputs, since the clock may not have been used in the description. Also, the data transfer at the interface may be treated in terms of, say, integer values, instead of logical values. In VHDL, each device model is treated as a distinct representation of a unique device, called an *entity* in this text. Figure 1-2 shows the VHDL view of a hardware device that has multiple device models, with each device model representing one entity. Even though entity 1 through *N* represent *N* different entities from the VHDL point of view, in reality they represent the same hardware device.

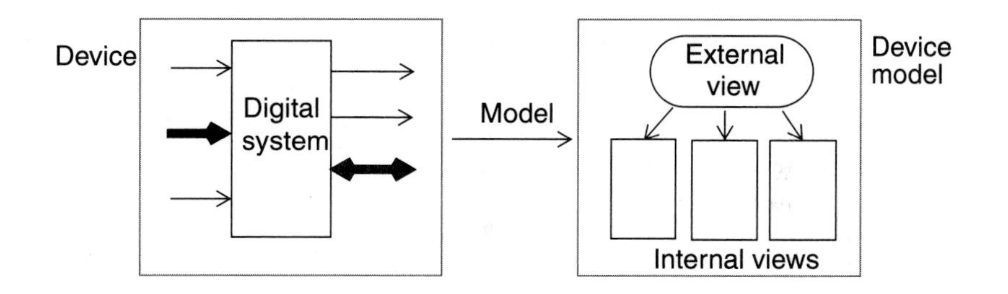

Figure 1-1 Device versus device model.

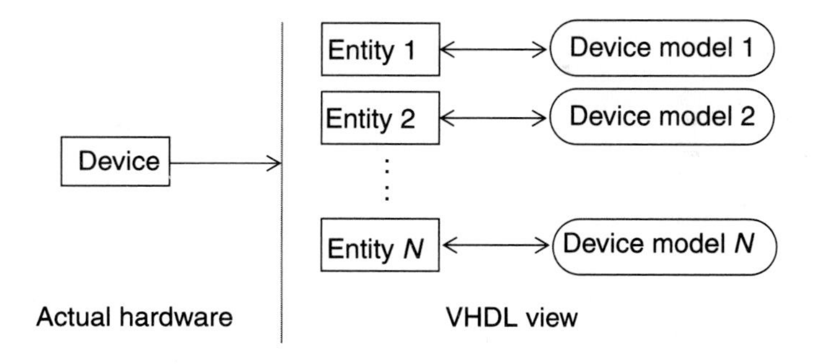

Figure 1-2 A VHDL view of a device.

The entity is thus a hardware abstraction of the actual hardware device. Each entity is described using one model, which contains one external view and one or more internal views. At the same time, a hardware device may be represented by one or more entities.

❑

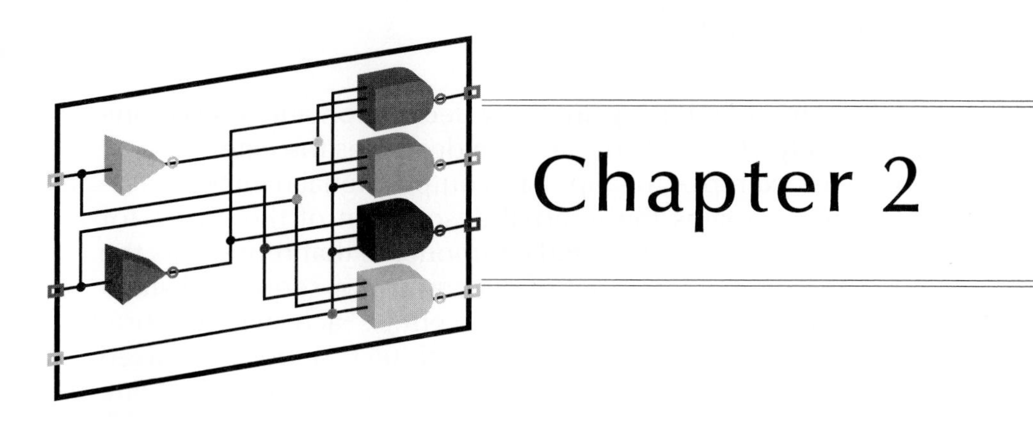

Chapter 2

A Tutorial

This chapter provides a quick introduction to the language. It describes the major modeling features of the language. At the conclusion of this chapter, you will be able to write simple VHDL models.

2.1 Basic Terminology

VHDL is a hardware description language that can be used to model a digital system. The digital system can be as simple as a logic gate or as complex as a complete electronic system. A hardware abstraction of this digital system is called an *entity* in this text. An entity X, when used in another entity Y, becomes a *component* for the entity Y. Therefore, a component is also an entity, depending on the level at which you are trying to model.

To describe an entity, VHDL provides five different types of primary constructs, called *design units*. They are:

1. Entity declaration
2. Architecture body
3. Configuration declaration

4. Package declaration
5. Package body

An entity is modeled using an entity declaration and at least one architecture body. The entity declaration describes the external view of the entity; for example, the input and output signal names. The architecture body contains the internal description of the entity; for example, as a set of interconnected components that represents the structure of the entity, or as a set of concurrent or sequential statements that represents the behavior of the entity. Each style of representation can be specified in a different architecture body or mixed within a single architecture body. Figure 2-1 shows an entity and one possible model.

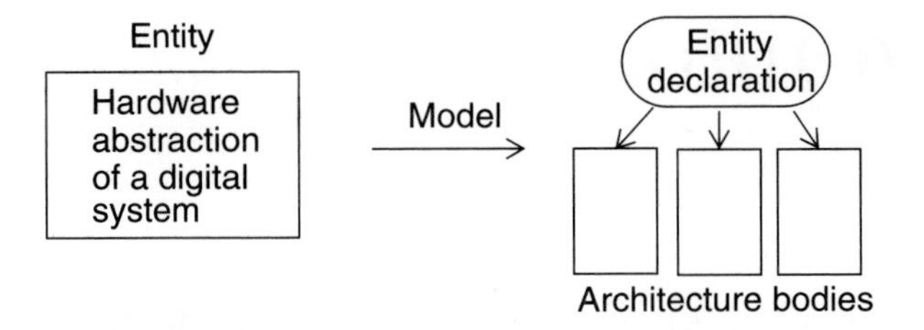

Figure 2-1 An entity and its model.

A configuration declaration is used to create a configuration for an entity. It specifies the binding of one architecture body from the many architecture bodies that may be associated with the entity. It may also specify the bindings of components used in the selected architecture body to other entities. An entity may have any number of different configurations.

A package declaration encapsulates a set of related declarations, such as type declarations, subtype declarations, and subprogram declarations, which can be shared across two or more design units. A package body contains the definitions of subprograms declared in a package declaration.

Figure 2-2 shows three entities called E1, E2, and E3. Entity E1 has three architecture bodies, E1_A1, E1_A2, and E1_A3. Architecture body E1_A1 is a purely behavioral model without any hierarchy. Architecture body E1_A2 uses a component called BX, while architecture body E1_A3 uses a component called CX. Entity E2 has two architecture bodies, E2_A1 and E2_A2, and architecture body E2_A1 uses a component called M1. Entity E3 has three architecture bodies, E3_A1, E3_A2, and E3_A3. Notice that each entity has a single entity declaration but more than one architecture body.

The dashed lines represent the binding that may be specified in a configuration for entity E1. There are two types of binding shown: binding of an architecture body to its entity and the binding of components used in the architecture body with other entities. For example, architecture body E1_A3 is bound to entity E1, while architecture body E2_A1 is bound to entity E2. Component M1 in architecture body E2_A1 is bound to entity E3. Component CX in the architecture body E1_A3 is bound to entity E2. However, one may choose a different configuration for entity E1 with the following bindings:

- Architecture E1_A2 is bound to its entity E1
- Component BX is bound to entity E3
- Architecture E3_A1 is bound to its entity E3

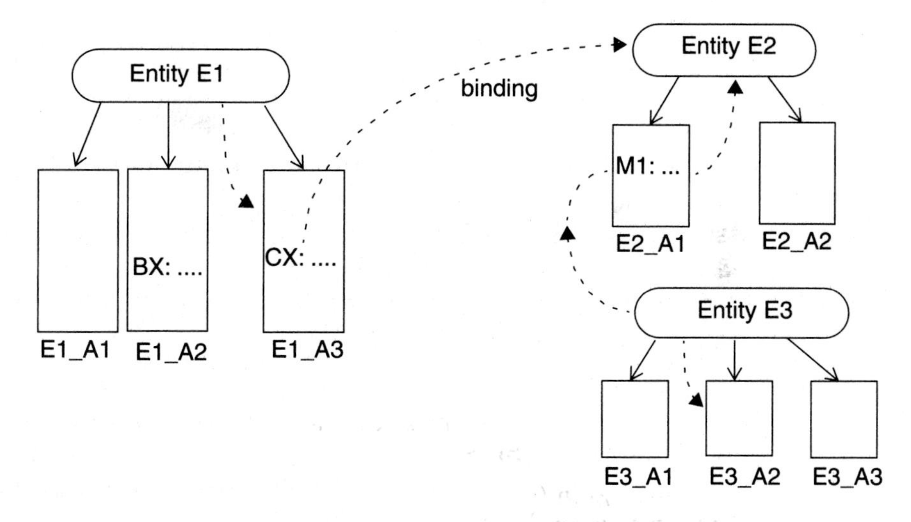

Figure 2-2 A configuration for entity E1.

Once an entity has been modeled, it needs to be validated by a VHDL system. A typical VHDL system consists of an analyzer and a simulator. The analyzer reads in one or more design units contained in a single file and compiles them into a design library after validating the syntax and performing some static semantic checks. The design library is a place in the host environment (that is, the environment that supports the VHDL system) where compiled design units are stored.

The simulator simulates an entity, represented by an entity-architecture pair or by a configuration, by reading in its compiled description from the design library and then performing the following steps:

1. Elaboration
2. Initialization
3. Simulation

A note on the language syntax. The language is case-insensitive; that is, lower-case and upper-case characters are treated alike (except in extended identifiers, string literals, and character literals). For example, CARRY, CarrY, or carrY all refer to the same name. The language is also free-format, very much like in Ada and Pascal. Comments are specified in the language by preceding the text with two consecutive dashes (--). All text between the two dashes and the end of that line is treated as a comment.

The terms introduced in this section are described in greater detail in the following sections.

2.2 Entity Declaration

The entity declaration specifies the name of the entity being modeled and lists the set of interface ports. *Ports* are signals through which the entity communicates with the other models in its external environment.

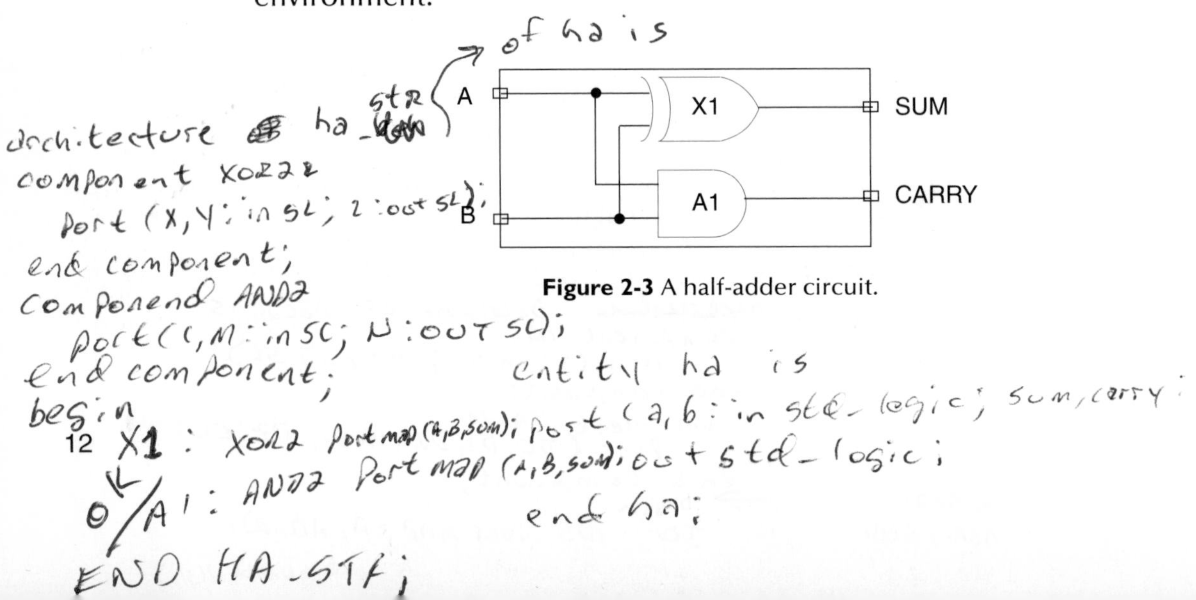

Figure 2-3 A half-adder circuit.

Here is an example of an entity declaration for the half-adder circuit shown in Figure 2-3.

```
entity HALF_ADDER is
   port (A, B: in BIT; SUM, CARRY: out BIT);
end HALF_ADDER;
-- This is a comment line.
```

The entity, called HALF_ADDER, has two input ports, A and B (the mode **in** specifies input port), and two output ports, SUM and CARRY (the mode **out** specifies output port). BIT is a predefined type of the language; it is an enumeration type containing the character literals '0' and '1'. The port types for this entity have been specified to be of type BIT, which means that the ports can take the values '0' or '1'.

The following is another example of an entity declaration for the 2-to-4 decoder circuit shown in Figure 2-4.

```
entity DECODER2x4 is
   port (A, B, ENABLE: in BIT; Z: out BIT_VECTOR(0 to 3));
end DECODER2x4;
```

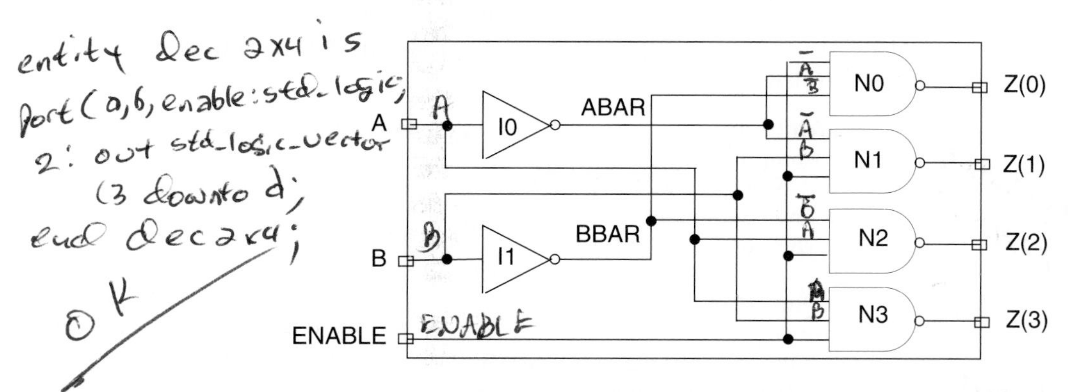

Figure 2-4 A 2-to-4 decoder circuit.

This entity, called DECODER2x4, has three input ports and four output ports. BIT_VECTOR is a predefined unconstrained array type of BIT. An unconstrained array type is a type in which the size of the array is not specified. The range "0 **to** 3" for port Z specifies the array size.

From the last two examples of entity declarations, we see that the entity declaration does not specify anything about the internals of the entity. It only specifies the name of the entity and the interface ports.

2.3 Architecture Body

The internal details of an entity are specified by an architecture body using any of the following modeling styles:

1. As a set of interconnected components (to represent structure)
2. As a set of concurrent assignment statements (to represent dataflow)
3. As a set of sequential assignment statements (to represent behavior)
4. As any combination of the above three

2.3.1 Structural Style of Modeling

In the structural style of modeling, an entity is described as a set of interconnected components. Such a model for the HALF_ADDER entity, shown in Figure 2-3, is described in an architecture body as shown below.

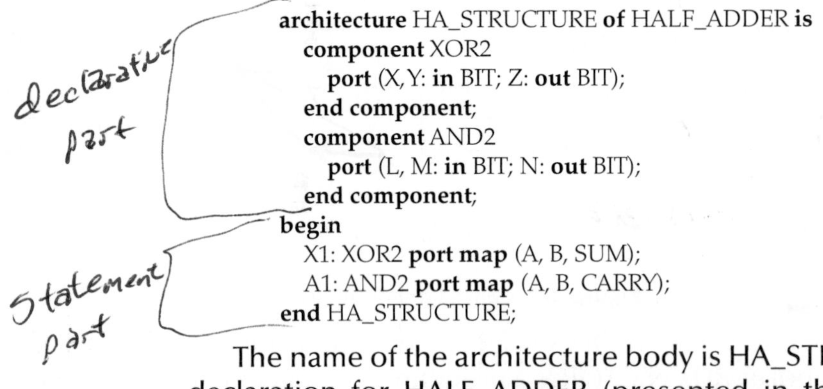

```
architecture HA_STRUCTURE of HALF_ADDER is
   component XOR2
      port (X,Y: in BIT; Z: out BIT);
   end component;
   component AND2
      port (L, M: in BIT; N: out BIT);
   end component;
begin
   X1: XOR2 port map (A, B, SUM);
   A1: AND2 port map (A, B, CARRY);
end HA_STRUCTURE;
```

The name of the architecture body is HA_STRUCTURE. The entity declaration for HALF_ADDER (presented in the previous section) specifies the interface ports for this architecture body. The architecture body is composed of two parts: the declarative part (before the keyword **begin**) and the statement part (after the keyword **begin**). Two component declarations are present in the declarative part of the architecture body. These declarations specify the interface of components that are used in the architecture body. The components XOR2 and AND2 may either be predefined components in a library or, if they do not exist, they may later be bound to other components in a library.

The declared components are instantiated in the statement part of the architecture body using component instantiation statements. X1 and A1 are the component labels for these component instantiations. The first component instantiation statement, labeled X1, shows that signals A and B (the input ports of the HALF_ADDER), are connected to the X and Y input ports of a XOR2 component, while output port Z of this component is connected to output port SUM of the HALF_ADDER entity.

Similarly, in the second component instantiation statement, signals A and B are connected to ports L and M of the AND2 component, while port N is connected to the CARRY port of the HALF_ADDER. Note that in this case, the signals in the port map of a component instantiation and the port signals in the component declaration are associated by position (called *positional association*). The structural representation for the HALF_ADDER does not say anything about its functionality. Separate entity models would be described for the components XOR2 and AND2, each having its own entity declaration and architecture body.

A structural representation for the DECODER2x4 entity, shown in Figure 2-4, is shown next.

```
architecture DEC_STR of DECODER2x4 is
  component INV
    port (PIN: in BIT; POUT: out BIT);
  end component;
  component NAND3
    port (D0, D1, D2: in BIT; DZ: out BIT);
  end component;
  signal ABAR, BBAR: BIT;
begin
  V0: INV port map (A, ABAR);
  V1: INV port map (B, BBAR);
  N0: NAND3 port map (ENABLE, ABAR, BBAR, Z(0));
  N1: NAND3 port map (ABAR, B, ENABLE, Z(1));
  N2: NAND3 port map (A, BBAR, ENABLE, Z(2));
  N3: NAND3 port map (A, B, ENABLE, Z(3));
end DEC_STR;
```

In this example, the name of the architecture body is DEC_STR, and it is associated with the entity declaration with the name

DECODER2x4; therefore, it inherits the list of interface ports from that entity declaration. In addition to the two component declarations (for INV and NAND3), the architecture body contains a signal declaration that declares two signals, ABAR and BBAR, of type BIT. These signals, which represent wires, are used to connect the various components that form the decoder. The scope of these signals is restricted to the architecture body; therefore, these signals are not visible outside the architecture body. Contrast these signals with the ports of an entity declaration, available for use within any architecture body associated with the entity declaration.

A component instantiation statement is a concurrent statement. Therefore, the order of these statements is not important. The structural style of modeling describes only an interconnection of components (viewed as black boxes), without implying any behavior of the components themselves nor of the entity that they collectively represent. In the architecture body DEC_STR, the signals A, B, and ENABLE, used in the component instantiation statements, are the input ports declared in the DECODER2x4 entity declaration. For example, in the component instantiation labeled N3, port A is connected to input D0 of component NAND3, port B is connected to input port D1 of component NAND3, port ENABLE is connected to input port D2, and port Z(3) of the DECODER2x4 entity is connected to the output port DZ of component NAND3. Again, positional association is used to map signals in a port map of a component instantiation with the ports of a component specified in its declaration. The behavior of the components NAND3 and INV are not apparent, nor is the behavior of the decoder entity that the structural model represents.

2.3.2 Dataflow Style of Modeling

In this modeling style, the flow of data through the entity is expressed primarily using concurrent signal assignment statements. The structure of the entity is not explicitly specified in this modeling style, but it can be implicitly deduced. Consider the following alternate architecture body for the HALF_ADDER entity that uses this style.

```
architecture HA_CONCURRENT of HALF_ADDER is
begin
   SUM <= A xor B after 8 ns;
   CARRY <= A and B after 4 ns;
end HA_CONCURRENT;
```

The dataflow model for the HALF_ADDER is described using two concurrent signal assignment statements (sequential signal assignment statements are described in the next section). In a signal assignment statement, the symbol <= implies an assignment of a value to a signal. The value of the expression on the right-hand-side of the statement is computed and is assigned to the signal on the left-hand-side, called the *target signal*. A concurrent signal assignment statement is executed only when any signal used in the expression on the right-hand-side has an event on it, that is, the value for the signal changes.

Delay information is included in the signal assignment statements using `after` clauses. If either signal A or B, which are input port signals of HALF_ADDER entity, has an event, say at time *T*, the right-hand-side expressions of both signal assignment statements are evaluated. Signal SUM is scheduled to get the new value after 8 ns, while signal CARRY is scheduled to get the new value after 4 ns. When simulation time advances to (*T*+4) ns, CARRY will get its new value, and when simulation time advances to (*T*+8) ns, SUM will get its new value. Thus, both signal assignment statements execute concurrently.

Concurrent signal assignment statements are concurrent statements, and therefore, the ordering of these statements in an architecture body is not important. Note again that this architecture body, with name HA_CONCURRENT, is also associated with the same HALF_ADDER entity declaration.

Here is a dataflow model for the DECODER2x4 entity.

```
architecture DEC_DATAFLOW of DECODER2x4 is
  signal ABAR, BBAR: BIT;
begin
  Z(3) <= not (A and B and ENABLE);              -- statement 1
  Z(0) <= not (ABAR and BBAR and ENABLE);        -- statement 2
  BBAR <= not B;                                 -- statement 3
  Z(2) <= not (A and BBAR and ENABLE);           -- statement 4
  ABAR <= not A;                                 -- statement 5
  Z(1) <= not (ABAR and B and ENABLE);           -- statement 6
end DEC_DATAFLOW;
```

The architecture body consists of one signal declaration and six concurrent signal assignment statements. The signal declaration declares signals ABAR and BBAR to be used locally within the architecture body. In each of the signal assignment statements, no `after` clause was used to specify delay. In all such cases, a default delay of 0 ns is assumed. This delay of 0 ns is also known as *delta delay*, and it represents an infinitesimally small delay.

To understand the behavior of this architecture body, consider an event happening on one of the input signals, say, input port B at time *T*. This would cause the concurrent signal assignment statements 1, 3, and 6 to be triggered. Their right-hand-side expressions would be evaluated, and the corresponding values would be scheduled to be assigned to the target signals at time ($T+\Delta$). When simulation time advances to ($T+\Delta$), new values are assigned to signals Z(3), BBAR, and Z(1). Since the value of BBAR changes, this will in turn trigger signal assignment statements 2 and 4. Eventually, at time ($T+2\Delta$), signals Z(0) and Z(2) will be assigned their new values.

The semantics of this concurrent behavior indicate that the simulation, as defined by the language, is event-triggered, and that simulation time advances to the next time unit when an event is scheduled to occur. Simulation time could also advance a multiple of delta time units. For example, events may have been scheduled to occur at times 1, 3, 4, $4+\Delta$, 5, 6, $6+\Delta$, $6+2\Delta$, $6+3\Delta$, 10, $10+\Delta$, 15, $15+\Delta$ time units.

The `after` clause may be used to generate a clock signal, as shown in the following concurrent signal assignment statement.

```
CLK <= not CLK after 10 ns;          -- Signal CLK is of type BIT.
```

This statement creates a periodic waveform on the signal CLK with a time period of 20 ns, as shown in Figure 2-5.

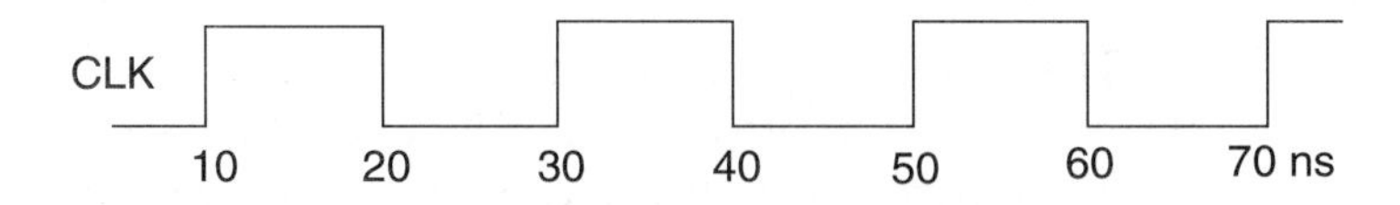

Figure 2-5 A clock waveform with constant on-off period.

2.3.3 Behavioral Style of Modeling

In contrast to the styles of modeling described earlier, the behavioral style of modeling specifies the behavior of an entity as a set of statements that are executed sequentially in the specified order. This set of sequential statements, which are specified inside a process statement, do not explicitly specify the structure of the entity but merely its functionality. A process statement is a concurrent statement that can appear within an architecture body. For example, consider the following behavioral model for the DECODER2x4 entity.

```
architecture DEC_SEQUENTIAL of DECODER2x4 is
begin
  process (A, B, ENABLE)
    variable ABAR, BBAR: BIT;
  begin
    ABAR := not A;                          -- statement 1
    BBAR := not B;                          -- statement 2
    if ENABLE = '1' then                    -- statement 3
      Z(3) <= not (A and B);                -- statement 4
      Z(0) <= not (ABAR and BBAR);          -- statement 5
      Z(2) <= not (A and BBAR);             -- statement 6
      Z(1) <= not (ABAR and B);             -- statement 7
    else
      Z <= "1111";                          -- statement 8
    end if;
  end process;
end DEC_SEQUENTIAL;
```

A process statement also has a declarative part (before the keyword **begin**) and a statement part (between the keywords **begin** and **end process**). The statements appearing within the statement part are sequential statements and are executed sequentially. The list of signals specified within the parentheses after the keyword **process** constitutes a sensitivity list, and the process statement is invoked whenever there is an event on any signal in this list. In the previous example, when an event occurs on signals A, B, or ENABLE, the statements appearing within the process statement are executed sequentially.

The variable declaration (starts with the keyword **variable**) declares two variables called ABAR and BBAR. A variable is different from a signal in that it is always assigned a value instantaneously, and the assignment operator used is the := compound symbol; contrast this with a signal that is assigned a value always after a certain delay (user-specified or the default delta delay), and the assignment operator used to assign a value to a signal is the <= compound symbol. Variables declared within a process have their scope limited to that process. Variables can also be declared in subprograms; subprograms are discussed in Chapter 8. Variables declared outside of a process or a subprogram are called *shared variables*. These variables can be updated and read by more than one process. Note that signals cannot be declared within a process.

Signal assignment statements appearing within a process are called *sequential signal assignment statements*. Sequential signal assignment statements, including variable assignment statements, are executed sequentially independent of whether an event occurs on any signals in its right-hand-side expression; contrast this with the execution of concurrent signal assignment statements in the dataflow modeling style. In the previous architecture body, if an event occurs on any signal, A, B, or ENABLE, statement 1, which is a variable assignment statement, is executed, then statement 2 is executed, and so on. Execution of the third statement, an `if` statement, causes control to jump to the appropriate branch based on the value of the signal ENABLE. If the value of ENABLE is '1', the next four signal assignment statements, 4 through 7, are executed independent of whether A, B, ABAR, or BBAR changed values, and the target signals are scheduled to get their respective values after delta delay. If ENABLE has a value '0', a value of '1' is assigned to each of the elements of the output array Z. When execution reaches the end of the process, the process suspends itself and waits for another event to occur on a signal in its sensitivity list.

It is possible to use `case` or `loop` statements within a process. The semantics and structure of these statements are very similar to those in other high-level programming languages like C or Pascal. An explicit `wait` statement can also be used to suspend a process. It can be used to wait for a certain amount of time, until a certain condition becomes true, or until an event occurs on one or more signals. Here is an example of a process statement that generates a clock with a different on-off period. Figure 2-6 shows the generated waveform.

```
process
begin
   CLK <= '0';
   wait for 20 ns;
   CLK <= '1';
   wait for 12 ns;
end process;
```

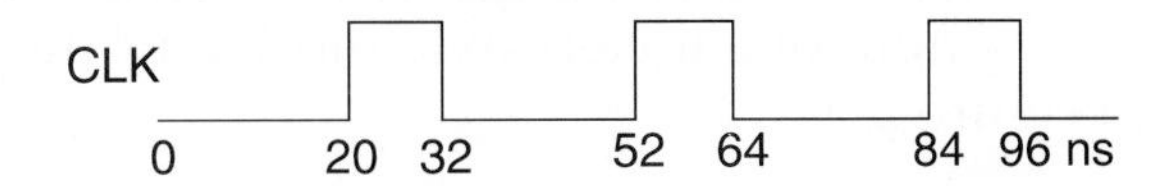

Figure 2-6 A clock waveform with varying on-off period.

This process does not have a sensitivity list because explicit `wait` statements are present inside the process. It is important to remember that a process never terminates. It is always either being executed or in a suspended state. All processes are executed once during the initialization phase of simulation until they get suspended. Therefore, a process with no sensitivity list and no explicit `wait` statements will never suspend itself.

A signal can represent not only a wire but also a place holder for a value, that is, it can be used to model a flip-flop. Here is such an example. Port signal Q models a level-sensitive flip-flop.

```
entity LS_DFF is
   port (Q: out BIT; D, CLK: in BIT);
end LS_DFF;

architecture LS_DFF_BEH of LS_DFF is
begin
   process (D, CLK)
   begin
      if CLK = '1' then
         Q <= D;
      end if;
   end process;
end LS_DFF_BEH;
```

The process executes whenever there is an event on signal D or CLK. If the value of CLK is '1', the value of D is assigned to Q. If CLK is '0', then no assignment to Q takes place. Thus, as long as CLK is '1', any change on D will appear on Q. Once CLK becomes '0', the value in Q is retained.

2.3.4 Mixed Style of Modeling

It is possible to mix the three modeling styles that we have seen so far in a single architecture body. That is, within an architecture body, we could use component instantiation statements (that represent structure), concurrent signal assignment statements (that represent dataflow), and process statements (that represent behavior). Here is an example of a mixed style model for the one-bit full-adder shown in Figure 2-7.

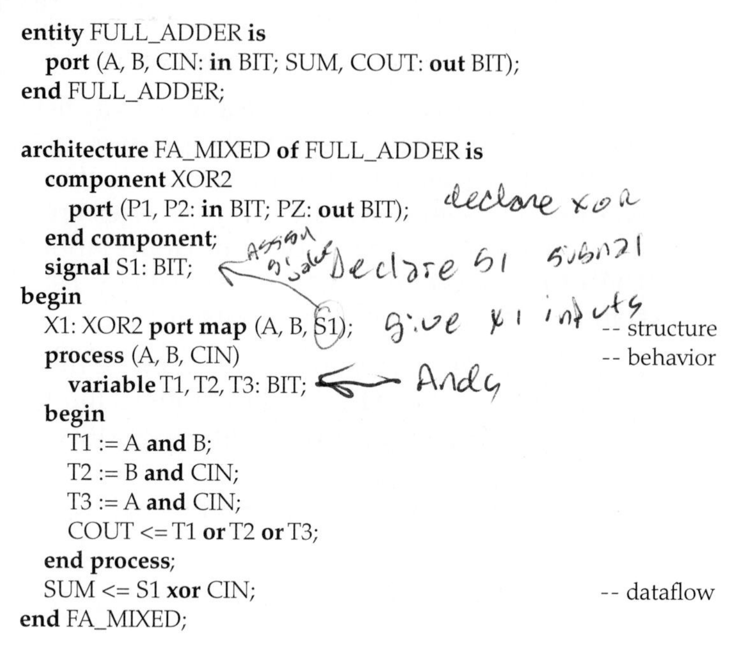

```
entity FULL_ADDER is
  port (A, B, CIN: in BIT; SUM, COUT: out BIT);
end FULL_ADDER;

architecture FA_MIXED of FULL_ADDER is
  component XOR2
    port (P1, P2: in BIT; PZ: out BIT);
  end component;
  signal S1: BIT;
begin
  X1: XOR2 port map (A, B, S1);                    -- structure
  process (A, B, CIN)                               -- behavior
    variable T1, T2, T3: BIT;
  begin
    T1 := A and B;
    T2 := B and CIN;
    T3 := A and CIN;
    COUT <= T1 or T2 or T3;
  end process;
  SUM <= S1 xor CIN;                                -- dataflow
end FA_MIXED;
```

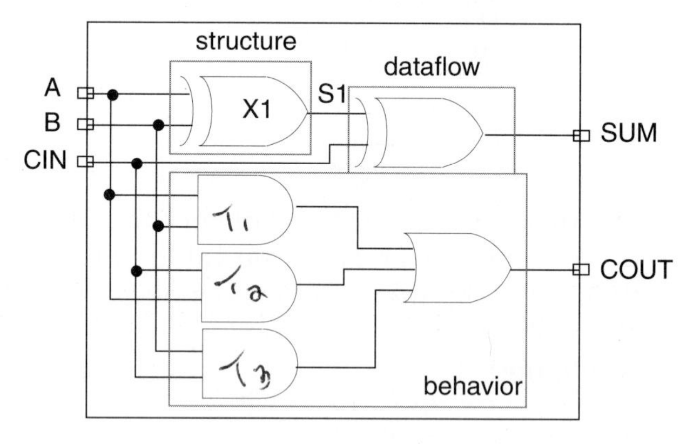

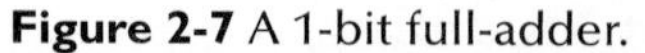
Figure 2-7 A 1-bit full-adder.

The full-adder is represented using one component instantiation statement, one process statement, and one concurrent signal assignment statement. All of these statements are concurrent statements; therefore, their order of appearance within the architecture body is not important. Note that a process statement itself is a concurrent statement; however, statements within a process statement are always executed sequentially. S1 is a signal locally declared within the architecture body and is used to pass the value from the output of the component X1 to the expression for signal SUM.

2.4 Configuration Declaration

A configuration declaration is used to select one of the possibly many architecture bodies that an entity may have, and to bind components, used to represent structure in that architecture body, to entities represented by an entity-architecture pair or by a configuration, which reside in a design library. Consider the following configuration declaration for the HALF_ADDER entity.

```
library CMOS_LIB, MY_LIB;
configuration HA_BINDING of HALF_ADDER is
  for HA_STRUCTURE
    for X1: XOR2
      use entity CMOS_LIB.XOR_GATE(DATAFLOW);
    end for;
    for A1: AND2
      use configuration MY_LIB.AND_CONFIG;
    end for;
  end for;
end HA_BINDING;
```

The first statement is a `library` clause that makes the library names CMOS_LIB and MY_LIB visible within the configuration declaration. The name of the configuration is HA_BINDING, and it specifies a configuration for the HALF_ADDER entity. The next statement specifies that the architecture body HA_STRUCTURE (described in Section 2.3.1) is selected for this configuration. Since this architecture body contains two component instantiations, two component bindings are required. The first statement (**for** X1: . . . **end for**) binds the component instantiation with label X1 to an entity represented by the entity-architecture pair, the XOR_GATE entity declaration, and the DATAFLOW architecture body, which resides in the CMOS_LIB design library. Similarly, component instantiation A1 is bound to a configuration of an entity defined by the configuration declaration, with name AND_CONFIG, residing in the MY_LIB design library.

There are no behavioral or simulation semantics associated with a configuration declaration. It merely specifies a binding that is used to build a configuration for an entity. These bindings are performed during the elaboration phase of simulation when the entire design to be simulated is being assembled. Having defined a configuration for the entity, the configuration can then be simulated.

An architecture body that does not contain any component instantiations, for example, when dataflow style is used, can also be selected to create a configuration. For example, the DEC_DATAFLOW architecture body can be selected for the DECODER2x4 entity using the following configuration declaration.

```
configuration DEC_CONFIG of DECODER2x4 is
  for DEC_DATAFLOW
  end for;
end DEC_CONFIG;
```

DEC_CONFIG defines a configuration that selects the DEC_DATAFLOW architecture body for the DECODER2x4 entity. The configuration DEC_CONFIG, which represents one possible configuration for the DECODER2x4 entity, can now be simulated.

2.5 Package Declaration

A package declaration is used to store a set of common declarations, such as components, types, procedures, and functions. These declarations can then be imported into other design units using a use clause. Here is an example of a package declaration.

```
package EXAMPLE_PACK is
  type SUMMER is (MAY, JUN, JUL, AUG, SEP);
  component D_FLIP_FLOP
    port (D, CK: in BIT; Q, QBAR: out BIT);
  end component;
  constant PIN2PIN_DELAY: TIME := 125 ns;
  function INT2BIT_VEC (INT_VALUE: INTEGER)
    return BIT_VECTOR;
end EXAMPLE_PACK;
```

The name of the package declared is EXAMPLE_PACK. It contains type, component, constant, and function declarations. Notice that the behavior of the function INT2BIT_VEC does not appear in the package declaration; only the function interface appears. The definition, or body, of the function appears in a package body (see the next section).

Assume that this package has been compiled into a design library called DESIGN_LIB. Consider the following clauses associated with an entity declaration.

```
library DESIGN_LIB;                       -- This is a library clause.
use DESIGN_LIB.EXAMPLE_PACK.all;          -- This is a use clause.
entity RX is . . .
```

The `library` clause makes the name of the design library DESIGN_LIB visible within this description; that is, the name DESIGN_LIB can be used within the description. This is followed by a `use` clause that imports all declarations in package EXAMPLE_PACK into the entity declaration of RX.

It is also possible to selectively import declarations from a package declaration into other design units. For example,

```
library DESIGN_LIB;
use DESIGN_LIB.EXAMPLE_PACK.D_FLIP_FLOP;
use DESIGN_LIB.EXAMPLE_PACK.PIN2PIN_DELAY;
architecture RX_STRUCTURE of RX is . . .
```

The two `use` clauses make the component declaration for D_FLIP_FLOP and the constant declaration for PIN2PIN_DELAY visible within the architecture body.

Another approach to selectively import items declared in a package is by using selected names. For example,

```
library DESIGN_LIB;         -- Package EXAMPLE_PACK has been
   -- compiled into a design library called DESIGN_LIB.
package ANOTHER_PACKAGE is
   function POCKET_MONEY
      (MONTH: DESIGN_LIB.EXAMPLE_PACK.SUMMER)
      return INTEGER;
   constant TOTAL_ALU: INTEGER;                -- A deferred constant.
end ANOTHER_PACKAGE;
```

The type SUMMER declared in package EXAMPLE_PACK is used in this new package by specifying a selected name. In this case, a `use` clause was not necessary. Package ANOTHER_PACKAGE also contains a constant declaration with the value of the constant not specified; such a constant is referred to as a *deferred constant*. The value of this constant is supplied in a corresponding package body.

See Appendix E for another example of a package declaration. This is the package STD_LOGIC_1164 which is an IEEE standard. Here is an example that uses this package.

```
library IEEE;                        -- Makes name IEEE visible.
use IEEE.STD_LOGIC_1164.all;         -- Makes all declarations,
  -- including types STD_LOGIC and STD_LOGIC_VECTOR,
  -- declared in package STD_LOGIC_1164 visible within
  -- entity declaration.
entity DECODER2x4 is
  port ( A, B, ENABLE: in STD_LOGIC;
         Z: out STD_LOGIC_VECTOR(0 to 3));
end DECODER2x4;

architecture DEC_DATAFLOW of DECODER2x4 is
  signal ABAR, BBAR: STD_LOGIC;
begin
  Z(3) <= not (A and B and ENABLE);
  Z(0) <= not (ABAR and BBAR and ENABLE);
  BBAR <= not B;
  Z(2) <= not (A and BBAR and ENABLE);
  ABAR <= not A;
  Z(1) <= not (ABAR and B and ENABLE);
end DEC_DATAFLOW;
```

The `library` clause declares the design library IEEE. The `use` clause imports all the declarations within the package STD_LOGIC_1164 into the entity declaration including the type declarations for STD_LOGIC and STD_LOGIC_VECTOR.

Appendix F describes a package UTILS_PKG that contains declarations and common utilities that are used in some examples of this book.

2.6 Package Body

A package body is used to store the definitions of functions and procedures that were declared in the corresponding package declaration, and also the complete constant declarations for any deferred constants that appear in the package declaration. Therefore, a package body is always associated with a package declaration. Furthermore, a package declaration can have at most one package body associated with it. Contrast this with an architecture body and an entity declaration, where multiple architecture bodies may be associated with a single entity declaration. A package body may contain other declarations as well (see Chapter 9).

Here is the package body for the package EXAMPLE_PACK, declared in the previous section.

```
package body EXAMPLE_PACK is
  function INT2BIT_VEC (INT_VALUE: INTEGER)
    return BIT_VECTOR is
  begin
    -- Behavior of function described here.
  end INT2BIT_VEC;
end EXAMPLE_PACK;
```

The name of the package body must be the same as that of the package declaration with which it is associated. It is important to note that a package body is not necessary if the corresponding package declaration has no function or procedure declarations and no deferred constant declarations. Here is the package body that is associated with the package ANOTHER_PACKAGE declared in the previous section.

```
package body ANOTHER_PACKAGE is
  -- A complete constant declaration:
  constant TOTAL_ALU: INTEGER := 10;
  function POCKET_MONEY                          -- Function body.
    (MONTH: DESIGN_LIB.EXAMPLE_PACK.SUMMER)
    return INTEGER is
  begin
    case MONTH is
      when MAY => return 5;
      when JUL | SEP => return 6;                -- When JUL or SEP.
      when others => return 2;                   -- When JUN or AUG.
    end case;
  end POCKET_MONEY;
end ANOTHER_PACKAGE;
```

2.7 Model Analysis

Once an entity is described in VHDL, it can be validated using an analyzer and a simulator that are part of a VHDL system. The first step in the validation process is analysis. The analyzer takes a file that contains one or more design units (remember that a design unit is an entity declaration, an architecture body, a configuration declaration, a package declaration, or a package body) and compiles them into an intermediate form. The format of this compiled intermediate representation is not defined by the language. During compilation, the analyzer validates the syntax and performs static semantic checks. The generated intermediate form is stored in a specific design library that has been designated as the working library.

A design library is a location in the host environment (the computer that supports the VHDL system) where compiled descriptions are stored. Each design library has a logical name that is used when referring to a library in a VHDL description. The mapping of the logical library name to a physical storage location is provided externally by the host environment and is not defined by the language. As an example, a design library can be implemented as a directory in the host environment with the compiled design units being stored as files within this directory. One possible way of providing the mapping of physical names to logical names is by specifying the mappings in a special file that the VHDL system can interpret.

An arbitrary number of design libraries may exist simultaneously. Of all the design libraries that may coexist, one particular library is designated as the working library with the logical name WORK. The language analyzer always compiles descriptions into this library; therefore, at any given time, only one library is updated. Figure 2-8 shows the compilation process.

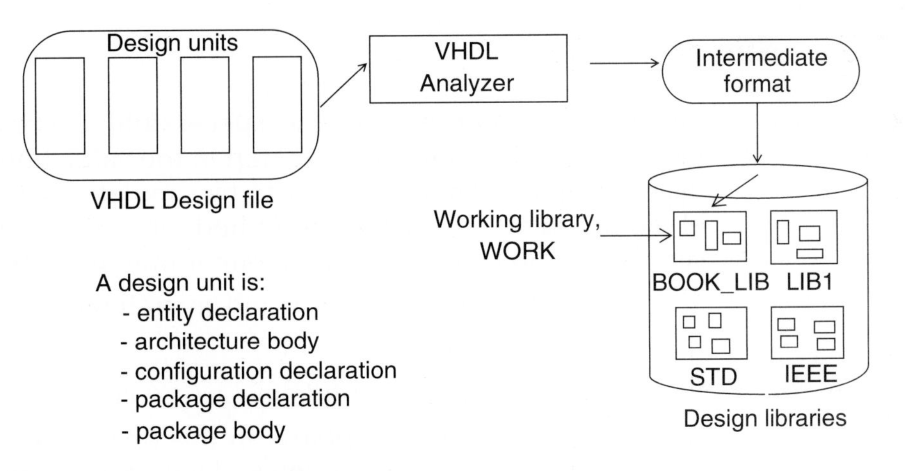

Figure 2-8 The compilation process.

If compiled descriptions need to be stored in a different design library, the reference to WORK is changed to point to this design library prior to analysis. The mapping of WORK to a design library is again provided by the host environment. Items compiled in one design library can be imported into design units compiled in a different design library by using `library` and `use` clauses, or by accessing them with a selected name.

There is a design library with the logical name *STD* predefined by the VHDL language environment. This library contains two packages: *STANDARD* and *TEXTIO*. The STANDARD package contains declarations for all the predefined types of the language (BIT, TIME, INTEGER, etc.). The TEXTIO package contains procedures and functions that are necessary for supporting formatted text read and write operations.

There also exists an IEEE standard package called *STD_LOGIC_1164*. This package defines a nine-value logic type, called *STD_ULOGIC*, and contains its associated subtypes, overloaded operator functions, and other useful utilities. This standard is called the IEEE Std 1164-1993. If available, a host environment must provide this package in a design library called *IEEE*.

2.8 Simulation

Once the model description is successfully compiled into one or more design libraries, the next step in the validation process is simulation. For a hierarchical entity to be simulated, all of its lowest-level components must be described at the behavioral level. A simulation can be performed on either one of the following:

- An entity declaration and an architecture body pair
- A configuration

Preceding the actual simulation are two major steps:

1. *Elaboration phase*: In this phase, the hierarchy of the entity is expanded and linked, components are bound to entities in a library, and the top-level entity is built as a network of behavioral models that is ready to be simulated. Also, storage is allocated for signals, variables, and constants declared in the design units. Initial values are also assigned to variables and constants. Files are opened if so indicated in their declarations.
2. *Initialization phase*: Driving and effective values for all explicitly declared signals are computed, implicit signals (discussed in later chapters) are assigned values, processes are executed once until they suspend, and simulation time is set to 0 ns.

Simulation commences by advancing time to that of the next event. Values that are scheduled to be assigned to signals at this time are assigned. If the value of a signal changes, and if that signal is present in the sensitivity list of a process, the process is executed until it suspends. Simulation stops when an assertion violation occurs, depending on the implementation of the VHDL system (assertion statements are discussed in Chapter 4) or when the maximum time as defined by the language is reached.

What to Expect Next

This chapter provided a brief overview of the major aspects of the language. Some of the other important features, like types, overloading, and resolution functions, were not discussed. The following chapters discuss these features and others in considerable detail.

❑

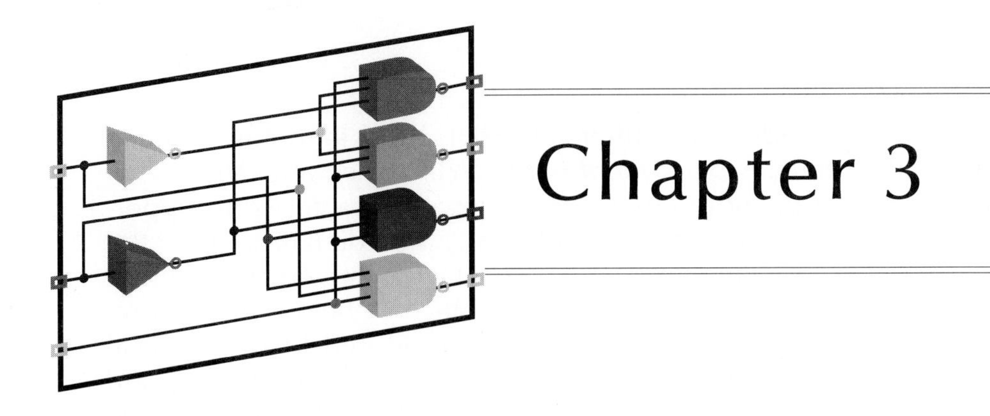

Chapter 3

Basic Language Elements

This chapter describes the basic elements of the language. These include data objects that store values of a given type, literals that represent constant values, and operators that operate on data values. Every data object belongs to a specific type. The various categories of types and the syntax for specifying user-defined types are discussed here. The chapter also describes how to associate types with objects by using object declarations.

It is important to understand the notion of data types and objects since VHDL is a strongly typed language. This means that operations and assignments are legal in the language only if the types of the operands and the result match according to a set of rules; it does not allow objects and literals of different types to be mixed freely in expressions. Examples of illegal operations are adding a real value to an integer value and assigning a Boolean value to an object of type BIT. It is therefore important to understand what types are and how they can be correctly used in the language.

A first-time reader may wish to skip the section on access types, incomplete types, and file types since this material is of a more advanced nature.

3.1 Identifiers

There are two kinds of identifiers in VHDL, basic identifiers and extended identifiers. A *basic identifier* in VHDL is composed of a sequence of one or more characters. A legal character is an upper-case letter (A . . . Z), a lower-case letter (a . . . z), a digit (0 . . . 9), or the underscore (_) character. The first character in a basic identifier must be a letter, and the last character may not be an underscore. Lower-case and upper-case letters are considered to be identical when used in a basic identifier; as an example, Count, COUNT, and CouNT all refer to the same basic identifier. Also, two underscore characters cannot appear consecutively. Some more examples of basic identifiers are:

```
DRIVE_BUS      SelectSignal   RAM_Address
SET_CK_HIGH    CONST32_59     r2d2
```

An *extended identifier* is a sequence of characters written between two backslashes. Any of the allowable characters can be used, including characters like ., !, @, ', and $. Within an extended identifier, lower-case and upper-case letters are considered to be distinct. Examples of extended identifiers are:

```
\TEST\                    -- Differs from the basic identifier TEST.
\-25\
\2FOR$\
\~Q\
\process\                 -- Distinct from the keyword process.
\.~$*****\
\7400TTL\
\----\\----\              -- Two consecutive backslashes represents
                          -- one backslash.
\COunt\          \COUNT\  -- Each is a distinct identifier and both are
                        -- distinct from the basic identifier COUNT.
```

Comments in a description must be preceded by two consecutive hyphens (--); the comment extends to the end of the line. Comments can appear anywhere within a description. Examples are:

```
-- This is a comment; it ends at the end of this line.
-- To continue a comment onto a second line, a separate
-- comment line must be started.

entity UART is end; -- This comment starts after entity declaration.
```

The language defines a set of reserved words; these are listed in Appendix A.1. These words, also called *keywords*, have a specific meaning in the language and therefore cannot be used as basic identifiers.

3.2 Data Objects

A data object holds a value of a specified type. It is created by means of an object declaration. An example is:

```
variable COUNT: INTEGER;
```

This results in the creation of a data object called COUNT, which can hold integer values. The object COUNT is also declared to be of *variable* class.

Every data object belongs to one of the following four classes:

1. *Constant*: An object of constant class (often called a *constant*) can hold a single value of a given type. This value is assigned to the constant before simulation starts, and the value cannot be changed during the course of the simulation. For a constant declared within a subprogram, the value is assigned to the constant every time the subprogram is called.
2. *Variable*: An object of variable class (often called a *variable*) can also hold a single value of a given type. However, in this case, different values can be assigned to the variable at different times using a variable assignment statement.
3. *Signal*: An object belonging to the signal class (often called a *signal*) holds a list of values, which include the current value of the signal, and a set of possible future values that are to appear on the signal. Future values can be assigned to a signal using a signal assignment statement.
4. *File*: An object belonging to the file class (often called a *file*) contains a sequence of values. Values can be read or written to the file using read procedures and write procedures, respectively.

Signals can be regarded as wires in a circuit, while variables and constants are analogous to their counterparts in a high-level programming language like C or Pascal. Signals are typically used to model wires and flip-flops, while variables and constants are typically used to model the behavior of the circuit. A file is used to model a file in the host environment.

An *object declaration* is used to declare an object, its type, and its class. A value can be optionally assigned to a signal, variable, or constant. For a file, the object declaration may specify information on how to open the file. Some examples of object declarations follow.

Constant Declarations

Examples of constant declarations are:

```
constant RISE_TIME: TIME := 10 ns;
constant BUS_WIDTH: INTEGER := 8;
```

The first declaration declares the object RISE_TIME, which can hold a value of type TIME (a predefined type in the language), and the value assigned to the object at the start of simulation is 10 ns. The second constant declaration declares a constant BUS_WIDTH of type INTEGER, with a value of 8.

An example of another form of constant declaration is:

```
constant NO_OF_INPUTS: INTEGER;
```

The value of the constant has not been specified in this case. Such a constant is called a *deferred constant*, and it can appear only inside a package declaration. The complete constant declaration with the associated value must appear in the corresponding package body.

Variable Declarations

Examples of variable declarations are:

```
variable CTRL_STATUS: BIT_VECTOR (10 downto 0);
variable SUM: INTEGER range 0 to 100 := 10;
variable FOUND, DONE: BOOLEAN;
```

The first declaration specifies a variable object CTRL_STATUS as an array of 11 elements, with each array element of type BIT. In the second declaration, an explicit initial value has been assigned to the variable SUM. When simulation starts, SUM will have an initial value of 10. If no initial value is specified for a variable object, a default value is used as an initial value. This default value is T'LEFT, where T is the object type and 'LEFT is a predefined attribute of a type that gives the leftmost value in the set of values belonging to type T. In the third declaration, the initial values assigned to FOUND and DONE at start of simulation is FALSE (FALSE is the leftmost value of the predefined type BOOLEAN). If the type of a variable is an array type or a record type, the initial value of each of the elements that form the variable is the initial value of each of its corresponding element types. Thus, the initial value for each of the array elements of CTRL_STATUS is '0'.

Signal Declarations

Here are some examples of signal declarations.

```
signal CLOCK: BIT;
signal DATA_BUS: BIT_VECTOR(0 to 7);
signal GATE_DELAY: TIME := 10 ns;
signal INIT_P: STD_LOGIC_VECTOR(7 downto 0) :=
                (0 => '1', others => 'U');
```

The interpretation of these signal declarations is very similar to that of the variable declarations. The first signal declaration declares the signal object CLOCK of type BIT and gives it an initial value of '0' ('0' being the leftmost value of type BIT). The third signal declaration declares a signal object GATE_DELAY of type TIME that has an initial value of 10 ns. The fourth signal declaration declares an 8-element signal array INIT_P of type STD_LOGIC_VECTOR; the 0th element is initialized to '1' and all other elements are initialized to the value 'U'. Other forms of signal declarations are described in Chapters 5 and 10.

File Declarations

A file is declared using a file declaration. The syntax of a file declaration is:

file *file-names* : *file-type-name* [[**open** *mode*] **is** *string-expression*] ;

The string expression is interpreted by the host environment as the physical name of a file. The mode specifies whether the file is to be used as a read-only or write-only, or in the append mode. Examples of file declarations are:

```
-- First two lines are file type declarations (explained later
-- in this chapter):
type STD_LOGIC_FILE is file of STD_LOGIC_VECTOR;
type BIT_FILE is file of BIT_VECTOR;

-- File declarations are:
file STIMULUS: TEXT open READ_MODE
                    is "/usr/home/jb/add.sti";
file VECTORS: BIT_FILE is "/usr/home/james/add.vec";
file PAT1, PAT2: STD_LOGIC_FILE;
```

In the first example of a file declaration, a file STIMULUS is declared to be of the predefined file type TEXT; that is, this file may contain an indefinite number of strings. The mode value, READ_MODE, specifies that the file will be opened in read-only mode, and the string expression after the keyword **is** specifies the path name to a physical file in the host environment. During elaboration, this declaration causes the specified file that STIMULUS is linked with "/usr/home/jb/add.sti", to be opened in read mode.

In the second example, VECTORS is declared as a file containing an indefinite number of bit vectors. This declaration also specifies the link to the file in the host environment. Since no mode is specified, the default mode, READ_MODE, is used. The third file declaration declares two files, PAT1 and PAT2, each of type STD_LOGIC_FILE; that is, the files contain values of type STD_LOGIC_VECTOR. Since no link has been specified to a file in the host environment, the file is not opened during elaboration. The file will have to be explicitly opened using a file open procedure.

Other Ways to Declare Objects

Not all objects in a VHDL description are explicitly created using object declarations. These other objects are declared as one of the following:

1. Ports of an entity. All ports are signal objects.
2. Generics of an entity (discussed in Chapter 7). These are constant objects.

3. Formal parameters of functions and procedures (discussed in Chapter 8). Function parameters are constants or signals, while procedure parameters can belong to any object class.

There are two other types of objects that are implicitly declared. These are the indices of a `for ... loop` statement and the `generate` statement (`generate` statements are discussed in Chapter 10). An example of such an implicit declaration for the loop index in a `for . . . loop` statement is shown.

```
for COUNT in 1 to 10 loop
  SUM := SUM + COUNT;
end loop;
```

In this `for ... loop` statement, COUNT is a constant that has an implicit declaration of type INTEGER with range 1 to 10, and therefore cannot be explicitly declared. The constant COUNT is created when the loop is first entered and ceases to exist after the loop is exited.

3.3 Data Types

Every data object in VHDL can hold a value that belongs to a set of values. This set of values is specified by using a *type declaration*. A *type* is a name that has associated with it a set of values and a set of operations. Certain types, and operations that can be performed on objects of these types, are predefined in the language. For example, INTEGER is a predefined type, with the set of values being integers in a specific range provided by the VHDL system. The minimum range that must be provided is $-(2^{31}-1)$ through $+(2^{31}-1)$. Some of the allowable and frequently used predefined operators are + for addition, – for subtraction, / for division, and * for multiplication. BOOLEAN is another predefined type that has the values FALSE and TRUE, and some of its predefined operators are **and**, **or**, **nor**, **nand**, **xor**, **xnor**, and **not**. The declarations for the predefined types of the language are contained in package STANDARD (see Appendix A); the operators for these types are predefined in the language.

The language also provides the facility to define new types by using type declarations and also to define a set of operations on these types by writing functions that return values of this new type. All the possible types that can exist in the language can be categorized into the following four major categories:

1. *Scalar* types: Values belonging to these types appear in sequential order.
2. *Composite* types: These are composed of elements of a single type (an array type) or elements of different types (a record type).
3. *Access* types: These provide access to objects of a given type (via pointers).
4. *File* types: These provide access to objects that contain a sequence of values of a given type.

It is possible to derive subtypes from predefined or user-defined types.

3.3.1 Subtypes

A *subtype* is a type with a constraint. The constraint specifies the subset of values for the subtype. The type is called the *base type* of the subtype. An object is said to belong to a subtype if it is of the base type and if it satisfies the constraint. *Subtype declarations* are used to declare subtypes. An object can be declared to belong to either a type or a subtype.

The set of operations belonging to a subtype is the same as that associated with its base type. Subtypes are useful for range checking and for imposing additional constraints on types.

Examples of subtypes are:

```
subtype MY_INTEGER is INTEGER range 48 to 156;
type DIGIT is ('0', '1', '2', '3', '4', '5', '6', '7', '8', '9');
  -- DIGIT is a type, not a subtype. It is the base type in the
  -- following subtype declaration:
subtype MIDDLE is DIGIT range '3' to '7';
```

In the first example, MY_INTEGER is a subtype of the INTEGER base type and has a range constraint with values ranging from 48 through 156. DIGIT is a user-defined enumeration type. The last subtype declaration declares a new subtype called MIDDLE, whose base type is DIGIT and has the values '3', '4', '5', '6', and '7'.

A subtype need not impose a constraint. In such a case, the subtype simply gives another name to an already existing type. For example,

```
subtype NUMBER is DIGIT;
```

NUMBER is a subtype with no constraint. Therefore, its set of values are the same as that for type DIGIT.

3.3.2 Scalar Types

The values belonging to these types are ordered, that is, relational operators can be used on these values. For example, BIT is a scalar type, and the expression '0' < '1' is valid and has the value TRUE. There are four different kinds of scalar types. These types are:

1. Enumeration
2. Integer
3. Physical
4. Floating point

Integer types, floating point types, and physical types are classified as *numeric* types because the values associated with these types are numeric. Enumeration and integer types are called *discrete* types because these types have discrete values associated with them. Every value belonging to an enumeration type, integer type, or a physical type has a *position number* associated with it. This number is the position of the value in the ordered list of values belonging to that type.

Enumeration Types

An enumeration type declaration defines a type that has a set of user-defined values consisting of identifiers and character literals. Examples are:

```
type MVL is ('U', '0', '1', 'Z');
type MICRO_OP is (LOAD, STORE, ADD, SUB, MUL, DIV);
subtype ARITH_OP is MICRO_OP range ADD to DIV;
```

Examples of objects defined for these types are:

```
signal CONTROL_A: MVL;
signal CLOCK: MVL range '0' to '1';     -- Implicit subtype declaration.
```

```
variable IC: MICRO_OP := STORE;
  -- STORE is the initial value for IC.
variable ALU: ARITH_OP;
```

MVL is an enumeration type that has the set of ordered values, 'U', '0', '1', and 'Z'. ARITH_OP is a subtype of the base type MICRO_OP and has a range constraint specified to be from ADD to DIV; that is, the values ADD, SUB, MUL, and DIV belong to the subtype ARITH_OP. A range constraint can also be specified in an object declaration, as shown in the signal declaration for CLOCK; here, the value of signal CLOCK is restricted to '0' or '1'.

The order in which values appear in an enumeration type declaration defines their ordering. That is, when using relational operators, a value is always less than a value that appears to its right in the order. For instance, in the MICRO_OP type declaration, STORE < DIV is true, and SUB > MUL is false. Values of an enumeration type also have a position number associated with them. The position number of the leftmost element is 0. The position number of any other element is one more than the position number of the element to its left.

The values of an enumeration type are called *enumeration literals*. For example, consider the following enumeration type declaration.

```
type CAR_STATE is (STOP, SLOW, MEDIUM, FAST);
```

The enumeration literals specified in this type declaration are STOP, SLOW, MEDIUM, and FAST; therefore, objects of type CAR_STATE may be assigned only these values.

If the same literal is used in two different enumeration type declarations, the literal is said to be *overloaded*. In such a case, whenever such a literal is used, the type of the literal is determined from its surrounding context. The following example clarifies this.

```
type MVL is ('U', '0', '1', 'Z');                -- line 1
type TWO_STATE is ('0', '1');                    -- line 2
. . .
variable CLOCK: TWO_STATE;                       -- line 3
variable LATCH_CTRL: MVL;                        -- line 4
. . .
CLOCK := '0';                                    -- line 5
LATCH_CTRL := LATCH_CTRL xor '0';                -- line 6
. . .
```

Here, '0' and '1' are two literals that are overloaded since they appear in both the types, MVL and TWO_STATE. The value '0' being assigned to CLOCK in line 5 refers to the literal in the TWO_STATE type since variable CLOCK is of that type. The value '0' in the expression for LATCH_CTRL refers to the literal in type MVL since the first argument of the **xor** operator is of type MVL (this assumes that the **xor** operator is allowed for MVL type operands, which is achieved by overloading the operator; operator overloading is discussed in Chapter 8).

The predefined enumeration types of the language are CHARACTER, BIT, BOOLEAN, SEVERITY_LEVEL, FILE_OPEN_KIND, and FILE_OPEN_STATUS. Values belonging to the type CHARACTER constitute the 191 characters of the ISO eight-bit coded character set. These values are called *character literals* and are always written between two single quotes (' '). Examples are:

```
'A'
'_'
'''     -- the single quote character itself.
'3'     -- the character literal 3.
```

The predefined type BIT has the literals '0' and '1', while type BOOLEAN has the literals FALSE and TRUE. Type SEVERITY_LEVEL has the values NOTE, WARNING, ERROR, and FAILURE; this type is typically used in assertion statements (assertion statements are described in Chapter 4). Type FILE_OPEN_KIND has the values READ_MODE, WRITE_MODE, and APPEND_MODE, while type FILE_OPEN_STATUS has the values OPEN_OK, STATUS_ERROR, NAME_ERROR, and MODE_ERROR; these two types are used exclusively with file operations.

Type STD_ULOGIC, defined in the package STD_LOGIC_1164, is an enumeration type. It is defined as:

```
type STD_ULOGIC is (
        'U',      -- Uninitialized
        'X',      -- Forcing unknown
        '0',      -- Forcing 0
        '1',      -- Forcing 1
        'Z',      -- High impedance
        'W',      -- Weak unknown
        'L',      -- Weak 0
        'H',      -- Weak 1
        '-'       -- Don't care
        );
```

Integer Types

An integer type defines a type whose set of values fall within a specified integer range. Examples of integer type declarations are:

```
type INDEX is range 0 to 15;
type WORD_LENGTH is range 31 downto 0;
subtype DATA_WORD is WORD_LENGTH range 15 downto 0;
type MY_WORD is range 4 to 6;
```

Some object declarations using these types are:

```
constant MUX_ADDRESS: INDEX := 5;
signal DATA_BUS: DATA_WORD;
```

INDEX is an integer type that includes the integer values from 0 through 15. DATA_WORD is a subtype of WORD_LENGTH that includes the integer values ranging from 15 through 0. The position of each value of an integer type is the value itself. For example, in the type declaration of WORD_LENGTH, value 31 is at position 31, value 14 is at position 14, and so on. In the declaration of MY_WORD, the position numbers of values 4, 5, and 6 are 4, 5, and 6, respectively. Contrast this with the position numbers of elements of an enumeration type; the position number in case of integer types does not refer to the index of the element in the range, but to its numeric value itself. The bounds of the range for an integer type must be constants or locally static expressions; a *locally static expression* is an expression that computes to a constant value at compile time (a *globally static expression* is an expression that computes to a constant value at elaboration time).

Values belonging to an integer type are called *integer literals*. Examples of integer literals are:

```
56349        6E2        0        98_71_28
```

Literal 6E2 refers to the decimal value $6 * (10^2) = 600$. The underscore (_) character can be used freely in writing integer literals and has no impact on the value of the literal; 98_71_28 is same as 987128.

INTEGER is the only predefined integer type of the language. The range of the INTEGER type is implementation-dependent but must at least cover the range from $-(2^{31} - 1)$ to $+(2^{31} - 1)$.

Floating Point Types

A floating point type has a set of values in a given range of real numbers. Examples of floating point type declarations are:

```
type TTL_VOLTAGE is range –5.5 to –1.4;
type REAL_DATA is range 0.0 to 31.9;
```

An example of an object declaration is:

```
variable LENGTH: REAL_DATA range 0.0 to 15.9;
...
variable L1, L2, L3: REAL_DATA range 0.0 to 15.9;
```

LENGTH is a variable object of type REAL_DATA that has been constrained to take real values in the range 0.0 through 15.9 only. Notice that in this case, the range constraint was specified in the variable declaration itself. Alternately, it is possible to declare a subtype and then use this subtype in the variable declarations as shown.

```
subtype RD16 is REAL_DATA range 0.0 to 15.9;
...
variable LENGTH: RD16;
...
variable L1, L2, L3: RD16;
```

The range bounds specified in a floating point type declaration must be constants or locally static expressions.

Floating point literals are values of a floating point type. Examples of floating point literals are:

```
16.26        0.0        0.002        3_1.4_2
```

Floating point literals differ from integer literals by the presence of the dot (.) character. Thus, 0 is an integer literal while 0.0 is a floating point literal.

Floating point literals can also be expressed in an exponential form. The exponent represents a power of 10 and the exponent value must be an integer. Examples are:

```
62.3E–2        5.0E+2
```

Integer and floating point literals can also be written in a base other than 10 (decimal). The base can be any value between 2 and 16. Such literals are called *based literals*. In this case, the exponent represents a power of the specified base. The syntax for a based literal is:

```
base # based-value #                    -- form 1
base # based-value # E exponent         -- form 2
```

Examples are:

2#101_101_000#	represents $(101101000)_2 = (360)$ in decimal,
16#FA#	represents $(FA)_{16} = (11111010)_2 = (250)$ in decimal,
16#E#E1	represents $(E)_{16} * (16^1) = 14 * 16 = (224)$ in decimal,
2#110.01#	represents $(110.01)_2 = (6.25)$ in decimal.

The base and the exponent values in a based literal must be in decimal notation.

The only predefined floating point type is REAL. The range of REAL is again implementation-dependent, but it must at least cover the range –1.0E38 to +1.0E38, and it must provide at least six decimal digits of precision.

Physical Types

A physical type contains values that represent measurement of some physical quantity, like time, length, voltage, or current. Values of this type are expressed as integer multiples of a base unit. An example of a physical type declaration is:

```
type CURRENT is range 0 to 1E9
  units
    nA;                         -- (base unit) nano-ampere
    uA  = 1000 nA;              -- micro-ampere
    mA  = 1000 uA;              -- milli-ampere
    Amp = 1000 mA;              -- ampere
  end units;

subtype FILTER_CURRENT is CURRENT range 10 uA to 5 mA;
```

CURRENT is defined to be a physical type that contains values from 0 nA to 10^9 nA. The base unit is a nano-ampere, while all others are derived units. The position number of a value is the number of base units represented by that value. For example, 2 uA has a position of 2000 while 100 nA has a position of 100. The range of values may include negative values, as shown in the next example.

```
type STEP_TYPE is range –10 to +10
  units
    STEP;                       -- base unit
    STEP2 = 2 STEP;             -- derived unit
    STEP5 = 5 STEP;             -- derived unit
  end units;
```

Values in this type range from –10 STEP to +10 STEP. Other units of this type are STEP2 and STEP5.

Values of a physical type are called *physical literals*. Physical literals are written as an integer or a floating point literal followed by the unit name. For example, "10 nA" is a physical literal (note that a space between 10 and nA is essential), while "Amp" is also a literal that implies 1 Amp. Other examples are:

```
100 ns
10V
50 sec
Kohm             -- implies 1 Kohm
5.2 mA           -- equivalent to 5200 uA
2.5 STEP2        -- equivalent to 5 STEP
5.6 nA           -- is 5 nA. Fractional part is truncated since nA is the
                 -- base unit for type CURRENT
7.2 STEP         -- is 7 STEP. Fractional part is truncated.
5.2643 uA        -- is 5264 nA
```

The only predefined physical type is TIME, and its range of base values, which again is implementation-dependent, must at least be $-(2^{31}-1)$ to $+(2^{31}-1)$. There is also a predefined physical subtype, DELAY_LENGTH, that represents non-negative time values. The declarations of type TIME and subtype DELAY_LENGTH appear in package STANDARD (see Appendix A).

3.3.3 Composite Types

A composite type represents a collection of values. There are two composite types: array types and record types. An array type represents a collection of values all belonging to a single type; on the other hand, a record type represents a collection of values that may belong to different types. An object belonging to a composite type therefore represents a collection of subobjects, one for each element of the composite type. A composite type could have a value belonging to either a scalar type, a composite type, or an access type. For example, a composite type may be defined to represent an array of an array of records. This provides the capability of defining arbitrarily complex composite types.

Array Types

An object of an array type consists of elements that have the same type. Examples of array type declarations are:

```
type ADDRESS_WORD is array (0 to 63) of BIT;
type DATA_WORD is array (7 downto 0) of STD_ULOGIC;
type ROM is array (0 to 125) of DATA_WORD;
type DECODE_MATRIX is array (POSITIVE range 15 downto 1,
   NATURAL range 3 downto 0) of STD_ULOGIC;
   -- POSITIVE and NATURAL are predefined subtypes; these are:
   subtype NATURAL is INTEGER range 0 to INTEGER'HIGH;
   subtype POSITIVE is INTEGER range 1 to INTEGER'HIGH;
   -- T 'HIGH gives the highest value belonging to type T.
```

Examples of object declarations using these types are:

```
variable ROM_ADDR: ROM;
signal ADDRESS_BUS: ADDRESS_WORD;
constant DECODER: DECODE_MATRIX;          -- A deferred constant.
variable DECODE_VALUE: DECODE_MATRIX;
```

ADDRESS_BUS is a one-dimensional array object that consists of 64 elements of type BIT. ROM_ADDR is a one-dimensional array object that consists of 126 elements, each element being another array object consisting of eight elements of type STD_ULOGIC. We have thus created an array of arrays.

Elements of an array can be accessed by specifying the index values into the array. For example, ADDRESS_BUS(26) refers to the 27th element of ADDRESS_BUS array object; ROM_ADDR(10)(5) refers to the value (of type STD_ULOGIC) at index 5 of the ROM_ADDR(10) data object (of type DATA_WORD); DECODER(5, 2) refers to the value of the element at the 2nd column and 5th row of the two-dimensional object. Notice the difference in addressing for a two-dimensional array type and an array of a type that is an array of another type.

The language allows for an arbitrary number of dimensions to be associated with an array. It also allows an array object to be assigned to another array object of the same type using a single assignment statement. Assignment can be made to an entire array, or to an element of an array, or to a slice of an array. For example,

```
ROM_ADDR(5) := "01000100";         -- Assign to an element of an array.
DECODE_VALUE := DECODER;           -- An entire array is assigned.
ADDRESS_BUS (8 to 15) <= X"FF";    -- Assign to a slice of an array.
```

These examples of array types are constrained array declarations, since the number of elements in the type is explicitly specified. The language also allows array types to be unconstrained; in this case, the number of elements in the array is not specified in the type declaration. Instead, an object declaration for an object of that type declares the number of elements of the array. A subtype declaration constraining the array type may also specify the index constraint for an unconstrained array type. A subprogram parameter may be of an unconstrained type; in this case, the constraint is obtained from the actual parameter passed in during the subprogram call. Subprograms are discussed in Chapter 8. Examples of unconstrained array declarations are:

```
-- The "<>" symbol is called "box".
type STACK_TYPE is array (INTEGER range <>)
  of ADDRESS_WORD;
subtype STACK is STACK_TYPE (0 to 63);
type OP_TYPE is (ADD, SUB, MUL, DIV);
type TIMING is array (OP_TYPE range <>, OP_TYPE range <>)
  of TIME;
```

Examples of object declarations using these types are:

```
variable FAST_STK: STACK_TYPE (–127 to 127);
constant ALU_TIMING: TIMING :=
  -- ADD, SUB, MUL
  ((10 ns, 20 ns, 45 ns),          -- ADD
   (20 ns, 15 ns, 40 ns),          -- SUB
   (45 ns, 40 ns, 30 ns));         -- MUL
```

STACK_TYPE is defined to be an unconstrained array type that specifies the index of the array as an integer type, and the element type as type ADDRESS_WORD. STACK is a subtype of the base type STACK_TYPE with the specified index constraint. The variable declaration for FAST_STK, defined to be of type STACK_TYPE, also specifies an index constraint. The constant ALU_TIMING specifies the timing for a two-operator ALU, where the operators could be ADD, SUB, or MUL; for example, an ALU that performs ADD and SUB operations has a delay of 20 ns. The declaration for ALU_TIMING is a special case of a constant declaration where no constraint need be specified for the unconstrained array type, since the size of the constant object is determined from the number of values in the constant.

There are two predefined one-dimensional unconstrained array types in the language, STRING and BIT_VECTOR. STRING is an array of characters, while BIT_VECTOR is an array of bits. Examples are:

```
variable MESSAGE: STRING(1 to 17) := "Hello,VHDL world";
signal RX_BUS: BIT_VECTOR(0 to 5) := O"37";
   -- O"37" is a bit-string literal representing the octal value 37.
constant ADD_CODE: BIT_VECTOR := ('0', '1', '1', '1', '0');
```

There are two one-dimensional unconstrained array types defined in the package STD_LOGIC_1164. These are:

```
type STD_ULOGIC_VECTOR is array (NATURAL range <>)
                                 of STD_ULOGIC;
type STD_LOGIC_VECTOR is array (NATURAL range <>)
                                 of STD_LOGIC;
   -- Type STD_LOGIC is a resolved subtype of STD_ULOGIC
   -- (see Section 5.4 for what a resolved subtype means).
```

Examples of declarations using these types are:

```
signal RASTER_LINE: STD_LOGIC_VECTOR(0 to NBITS);
variable REG_FILE: STD_ULOGIC_VECTOR
                        (UPPER_BND downto LOWER_BND);
```

VHDL does not allow a type that is an unconstrained array of an unconstrained array.

```
type MEMORY is array (NATURAL range <>) of
            STD_LOGIC_VECTOR;          -- Not allowed in VHDL.

type REG_FILE is array (NATURAL range <>) of
     BIT_VECTOR(0 to 7);               -- This is ok.
type DISPLAY_BANK is
     array (NATURAL range <>, NATURAL range <>) of BIT;
                                       -- This is also ok.
```

The first type declaration, that of MEMORY, is not valid VHDL since STD_LOGIC_VECTOR is an unconstrained array type and therefore MEMORY is an unconstrained array of an unconstrained array, which is not allowed. Unconstrained arrays of constrained arrays are allowed as the example of type REG_FILE shows. The third type declaration, that of DISPLAY_BANK, shows that a two-dimensional (including multi-dimensional) unconstrained array type is supported by the language.

A value representing a one-dimensional array of characters is called a *string literal*. String literals are written by enclosing the sequence of characters within double quotes ("..."). Examples of string literals are:

```
"THIS IS A TEST"
"SPIKE DETECTED!"
"State ""READY"" entered!"     -- Two double quote characters in a
                -- sequence represents one double quote character.
```

A string literal can be assigned to different types of objects; for example, to a STRING type object or a BIT_VECTOR type object. The type of a string literal is therefore determined from the context in which it appears. Here are some examples.

```
-- Example 1:
variable ERROR_MESSAGE: STRING(1 to 19);
ERROR_MESSAGE := "Fatal ERROR: abort!";

-- Example 2:
variable BUS_VALUE: BIT_VECTOR(0 to 3);
BUS_VALUE := "1101";
```

In the first example, the string literal is of type STRING, while in the second example, the string literal is of type BIT_VECTOR. The type of a string literal can also be explicitly stated by using a qualified expression (qualified expressions are described in Chapter 10).

A string literal that represents a sequence of bits (values of type BIT) can also be represented as a *bit string literal*. This sequence of bits, called *bit strings*, can be represented as either a binary value, an octal value, or a hexadecimal value. The underscore character can be freely used in bit string literals for clarity. Examples are:

```
X"FF0"                  -- X for hexadecimal.
B"00_0011_1101"         -- B for binary.
O"327"                  -- O for octal.
X"FF_F0_AB"
```

The type of a bit string literal is also determined from the context in which it appears. For example,

```
type MVL is ('X', '0', '1', 'Z');
type MVL_VECTOR is array (NATURAL range <>) of MVL;
variable FXA: MVL_VECTOR (0 to 3);
variable BRY: BIT_VECTOR (7 downto 0);
variable CLA: STD_LOGIC_VECTOR (1 to 9);
...
FXA := B"01_00";          -- Assign a binary value.
```

```
BRY := X"AB";            -- Assign a hexadecimal value.
CLA := O"304";           -- Assign an octal value.
```

The bit string literal in the first assignment statement is of type MVL_VECTOR, the bit string literal in the second assignment statement is of type BIT_VECTOR, while the bit string literal in the third assignment is of type STD_LOGIC_VECTOR.

There are many different ways to assign values to an array object. Here are some examples.

```
variable OP_CODES : BIT_VECTOR(1 to 5);
OP_CODES := "01001";                      -- A string literal is assigned.
OP_CODES := ('0', '1', '0', '0', '1');    -- Positional association is used;
  -- first value is assigned to OP_CODES(1), second value to
  -- OP_CODES(2), and so on.
OP_CODES := (2 => '1', 5 => '1', others => '0'); -- Named association;
  -- second and fifth element of OP_CODES get the value '1', and
  -- the rest get a value of '0'.
OP_CODES := (others => '0');              -- All values set to '0'.
```

Notice the use of the keyword **others** to mean all previously unassigned values; when used, it must be the last association. The expressions used in the last three assignments to OP_CODES are examples of array aggregates. An *aggregate* is a set of comma-separated elements enclosed within parenthesis.

An aggregate can also be used to provide an initial value for an array object in its declaration. Here are two examples.

```
variable OP_CODES : BIT_VECTOR(1 to 5) :=
                          (2 | 5 => '1', others => '0');
signal CG_PORT: STD_LOGIC_VECTOR(0 to 15) := (others => '0');
```

In the variable declaration, the 2nd and the 5th bits of the variable OP_CODES are set to '1', while all other bits are set to '0'. In the signal declaration, all elements of the array CG_PORT are initialized to the value '0'.

Here is an example of an aggregate specified in a constant declaration.

```
constant CLN: STD_LOGIC_VECTOR (2 downto 0) := "001";
```

CLN(2) is '0', CLN(1) is '0' and CLN(0) is '1'. When an aggregate has only one element, then named association must be used to specify the value.

```
constant TIK: STD_LOGIC_VECTOR (2 downto 2) := (2 => '1');
-- It is illegal to write:
-- constant TIK: STD_LOGIC_VECTOR(2 downto 2) := ('1');
```

Record Types

An object of a record type is composed of elements of same or different types. A record type is analogous to the `record` data type in Pascal and the `struct` declaration in C. An example of a record type declaration is:

```
type PIN_TYPE is range 0 to 10;
type MODULE is
  record
    SIZE              : INTEGER range 20 to 200;
    CRITICAL_DLY      : TIME;
    NO_INPUTS         : PIN_TYPE;
    NO_OUTPUTS        : PIN_TYPE;
  end record;
```

Values can be assigned to a record object using aggregates. For example,

```
variable NAND_COMP: MODULE;
  -- NAND_COMP is an object of record type MODULE.
NAND_COMP := (50, 20 ns, 3, 2);
  -- Implies 50 is assigned to SIZE, 20 ns is assigned
  -- to CRITICAL_DLY, etc.
```

Values can be assigned to a record object from another record object of the same type using a single assignment statement. In the following example, each element of NAND_GENERIC is assigned the value of the corresponding element in NAND_COMP.

```
signal NAND_GENERIC: MODULE;

NAND_GENERIC <= NAND_COMP;
```

Each element of a record object can also be assigned individually by using selected names. For example,

```
NAND_COMP.NO_INPUTS := 2;
```

Aggregate values can be assigned to record objects using both positional and named associations. Since the type of an aggregate is determined from the context in which it is used, the aggregate can be either an array or a record aggregate, depending on its usage. For example,

```
CAB := (others => '0');
```

If CAB is an array type object, the aggregate is treated as an array aggregate; on the other hand, if CAB is a record type object, the aggregate is a record aggregate where **others** refers to all the elements in the record.

3.3.4 Access Types

Values belonging to an access type are pointers to a dynamically allocated object of some other type. They are similar to pointers in Pascal and C languages. Examples of access type declarations are:

```
-- MODULE is a record type declared in the previous sub-section.
type PTR is access MODULE;
type FIFO is array (0 to 63, 0 to 7) of BIT;
type FIFO_PTR is access FIFO;
```

PTR is an access type whose values are addresses that point to objects of type MODULE. Every access type may also have the value **null,** which means that it does not point to any object. Objects of an access type can only belong to the `variable` class. When an object of an access type is declared, the default value of this object is **null.** For example,

```
variable MOD1PTR, MOD2PTR: PTR;          -- Default value is null.
```

Objects to which access types point can be created using *allocators.* Allocators provide a mechanism to dynamically create objects of a specific type.

```
MOD1PTR := new MODULE;
```

The **new** in this assignment causes an object of type MODULE to be created, and the pointer to this object is returned. The values of the elements of the MODULE record are the default values of each element; these are 20 (the leftmost value of the implied subtype) for the SIZE element, TIME'LEFT for the CRITICAL_DLY element, and the value 0 (this is PIN_TYPE'LEFT) for the NO_INPUTS and NO_OUTPUTS elements. Initial values can also be assigned to a newly created object by explicitly specifying the values, as shown in the following example.

```
MOD2PTR := new MODULE'(25, 10 ns, 4, 9);
```

Objects of an access type can be referenced as:

1. *obj-ptr*.**all**: Accesses the entire object pointed to by obj_ptr, where *obj-ptr* is a pointer to an object of any type
2. *array-obj-ptr* (*element-index*): Accesses the specified array element, where *array-obj-ptr* is a pointer to an array object
3. *record-obj-ptr.element-name*: Accesses the specified record element, where *record-obj-ptr* is a pointer to a record object

The elements of the object to which MOD2PTR points can be accessed as MOD2PTR.SIZE, MOD2PTR.CRITICAL_DLY, MOD2PTR.NO_INPUTS, and MOD2PTR.NO_OUTPUTS, provided the pointer is not null.

For every access type, a procedure DEALLOCATE is implicitly declared. This procedure, when called, returns the storage occupied by the object to the host environment. For the PTR and FIFO_PTR access types declared earlier, the following DEALLOCATE procedures are implicitly declared:

```
procedure DEALLOCATE (P: inout PTR);
procedure DEALLOCATE (P: inout FIFO_PTR);
```

The execution of the following statement:

```
DEALLOCATE(MOD2PTR);
```

causes the storage occupied by the object to which MOD2PTR points, to be deallocated, and MOD2PTR is set to **null**.

Pointers can be assigned to other pointer variables of the same access type. Therefore,

```
MOD1PTR := MOD2PTR;
```

is a legal assignment. Both MOD1PTR and MOD2PTR now point to the same object. Consider the following example.

```
type BITVEC_PTR is access BIT_VECTOR;
variable BITVEC1: BITVEC_PTR := new BIT_VECTOR'("1001");
```

BITVEC1 points to an object that is constrained to be of four bits. The bits can be accessed as BITVEC1(0), which is 1, BITVEC1(1), which is 0, and so on. Here is another example in which values are assigned to a dynamically allocated object using named association.

```
MOD2PTR := new MODULE'(CRITICAL_DLY => 10 ns,
                NO_INPUTS => 2, NO_OUTPUTS => 3, SIZE => 100);
```

Access types are useful in modeling high-level behavior, especially that of regular structures like RAMs and FIFOs, where pointers can be used to access objects one at a time in a sequence.

3.3.5 Incomplete Types

It is possible to have an access type that points to an object which has elements that are also access types. This can lead to mutually dependent or recursive access types. Since a type must be declared before it is used, an incomplete type declaration can be used to solve this problem. An incomplete type declaration has the form:

type *type-name* ;

Once an incomplete type has been declared, the *type-name* can now be used in any mutually dependent or recursive access type. However, a corresponding full type declaration must follow later. An example of a mutually dependent access type is:

```
type COMP;          -- Record contains component name and list of
                    -- nets its connected to.
type NET;           -- Record contains net name and list of
                    -- components it is connected to.
type COMP_PTR is access COMP;
type NET_PTR is access NET;
constant MODMAX: INTEGER := 100;
constant NETMAX: INTEGER := 2500;
type COMP_LIST is array (1 to MODMAX) of COMP_PTR;
type NET_LIST is array (1 to NETMAX) of NET_PTR;

type COMPLIST_PTR is access COMP_LIST;
type NETLIST_PTR is access NET_LIST;

-- Full type declarations for COMP and NET follow:
type COMP is
  record
    COMP_NAME        : STRING(1 to 10);
    NETS             : NETLIST_PTR;
  end record;

type NET is
  record
    NET_NAME           : STRING(1 to 10);
    COMPONENTS         : COMPLIST_PTR;
  end record;
```

Here, COMP and NET have elements that access objects of type NET and COMP, respectively. An example of a recursive access type is:

```
type DFG;                                    -- A data flow graph node.
type OP_TYPE is (ADD, SUB, MUL, DIV, SHIFT, ROTATE);
type PTR is access DFG;
type DFG is
  record
    OP_CODE      : OP_TYPE;
    SUCC         : PTR;                      -- Successor node list.
    PRED         : PTR;                      -- Predecessor node list.
  end record;
```

PTR is an access type of DFG. It is also the type of an element in DFG. Here is another example.

```
type MEM_RANGE is range 0 to 16383;
type HOLE;                           -- Represents available memory.
type HOLE_PTR is access HOLE;
type HOLE is
  record
    START_LOC, END_LOC               : MEM_RANGE;
    PREV_HOLE, NEXT_HOLE             : HOLE_PTR;
  end record;
```

3.3.6 File Types

Objects of file types represent files in the host environment. They provide a mechanism by which a VHDL design communicates with the host environment. The syntax of a file type declaration is:

type *file-type-name* **is file of** *type-name*;

The *type-name* is the type of values contained in the file. Here are two examples.

```
type VECTORS is file of BIT_VECTOR;
type NAMES is file of STRING;
```

A file of type VECTORS has a sequence of values of type BIT_VECTOR; a file of type NAMES has a sequence of strings as values in it.

A file can be opened, closed, read, written to, or tested for an end-of-file condition by using special procedures and functions that are implicitly declared for every file type. These are:

```
procedure FILE_OPEN (file F: file-type-name;
        EXTERNAL_NAME: in STRING;
        OPEN_KIND: in FILE_OPEN_KIND := READ_MODE);
  -- Opens the file F that points to the physical file specified in the
  -- string EXTERNAL_NAME, with the specified mode.
  -- Type FILE_OPEN_KIND has values:
  -- READ_MODE (default), WRITE_MODE, and APPEND_MODE.

procedure FILE_OPEN (STATUS: out FILE_OPEN_STATUS;
        file F: file-type-name;
        EXTERNAL_NAME: in STRING;
        OPEN_KIND: in FILE_OPEN_KIND := READ_MODE);
  -- Same as first procedure, except that this procedure returns the file
  -- open status. Type FILE_OPEN_STATUS has the following values:
  -- OPEN_OK            : file open was successful.
  -- STATUS_ERROR       : file was already open.
  -- NAME_ERROR         : file not found or not accessible.
  -- MODE_ERROR         : could not open file with specified
                        : access mode.

procedure FILE_CLOSE (file F: file-type-name);
  -- Closes the specified file.

procedure READ (file F: file-type-name; VALUE: out type-name);
  -- Gets the next value in VALUE from file F.

procedure WRITE (file F: file-type-name; VALUE: in type-name);
  -- Appends a given value in VALUE to file F.

function ENDFILE (file F: file-type-name) return BOOLEAN;
  -- Returns false if a read on a file F will be successful in getting
  -- another value, otherwise it returns true.
```

If *type-name* is an unconstrained array type, a different READ procedure is implicitly declared, which is of the form:

```
procedure READ (file F: file-type-name; VALUE: out type-name;
          LENGTH: out NATURAL);
  -- LENGTH returns the number of elements of the array that
  -- was read.
```

Values within a file can only be accessed sequentially. A file can be opened implicitly by specifying the file open information in the file declaration. Some examples are shown next.

```
file DUMP: NAMES open APPEND_MODE is "top.dump";
  -- An implicit call to FILE_OPEN is performed at elaboration time:
  -- FILE_OPEN (DUMP, "top.dump", APPEND_MODE);

file PATTERNS: VECTORS is "uart.pat";
  -- An implicit call to FILE_OPEN is performed at elaboration time:
  -- FILE_OPEN (PATTERNS, "uart.pat", READ_MODE);
```

```
file TMP: VECTORS;
   -- Since no file open information is provided, an explicit call to
   -- FILE_OPEN needs to be performed before the file TMP can be
   -- accessed.
```

Here is a complete example of a test bench that reads vectors from a file, "fadd.vec", applies these vectors to the test component, a one-bit full adder, and then writes the results into another file, "fadd.out".

```
entity FA_TEST is end;

architecture IO_EXAMPLE of FA_TEST is
   component FULL_ADD
      port (CIN, A, B: in BIT; COUT, SUM: out BIT);
   end component;
   subtype STRING3 is BIT_VECTOR(0 to 2);
   subtype STRING2 is BIT_VECTOR(0 to 1);
   type IN_TYPE is file of STRING3;
   type OUT_TYPE is file of STRING2;

   file VEC_FILE: IN_TYPE
      open READ_MODE is "/usr/home/jb/vhdl_ex/fadd.vec";
   file RESULT_FILE: OUT_TYPE
      open WRITE_MODE is "/usr/home/jb/vhdl_ex/fadd.out";
   signal S: STRING3;
   signal Q: STRING2;
begin
   FA: FULL_ADD port map (S(0), S(1), S(2), Q(0), Q(1));

   process
      constant PROPAGATION_DELAY: TIME := 25 ns;
      variable IN_STR: STRING3;
      variable OUT_STR: STRING2;
   begin
      while not ENDFILE (VEC_FILE) loop
         READ (VEC_FILE, IN_STR);
         S <= IN_STR;
         wait for PROPAGATION_DELAY;
         OUT_STR := Q;
         WRITE (RESULT_FILE, OUT_STR);
      end loop;
      report "Completed processing all vectors.";
      wait;                                    -- Stop simulation.
   end process;
end IO_EXAMPLE;
```

Two files, VEC_FILE and RESULT_FILE, are declared. VEC_FILE is an input file and contains three-bit strings, and RESULT_FILE is an output file into which two-bit strings are written. Input vectors are read one at a time until end of file is reached. Each vector is applied, and then the process waits for the full-adder circuit to stabilize before sampling the adder output value which is written to the output file. The report statement, when executed, simply prints the report message. The wait statement causes the process to suspend indefinitely.

One file type, TEXT, is predefined in the language; this file type represents a file consisting of variable-length text strings.

```
type TEXT is file of STRING;
  -- A file of variable-length strings.
```

An access type, LINE, is provided to point to such strings.

```
type LINE is access STRING;        -- A line is a pointer to a STRING.
```

Operations to read and write data from a single line are provided. Operations to read and write entire lines are also provided. The definitions for all these types and operations appear in the predefined package, TEXTIO.

Here is an example that reads from each line, a last name (a string), an ID number (an integer) and a dollar amount (a real number), and rewrites out to another file in the reverse order, that is, dollar amount first, ID second and then the last name.

```
use STD.TEXTIO.all;        -- Include the predefined package TEXTIO.
entity REVERSE_RECORDS is end;

architecture SIMPLE of REVERSE_RECORDS is
  file FROM_FILE: TEXT open READ_MODE
                      is "/home/jb/vhdl_ex/master_records.in";
  file TO_FILE: TEXT open WRITE_MODE
                  is "/home/jb/vhdl_ex/master_records.out";
begin
  process
    variable BUF_IN, BUF_OUT: LINE;
    variable LNAME: STRING(1 to 20);
    variable ID_NUM: INTEGER;
    variable DOLLARS: REAL;
  begin
    while not ENDFILE(FROM_FILE) loop
      READLINE (FROM_FILE, BUF_IN);              -- Read line from file.
```

```
      READ (BUF_IN, LNAME);                -- Read string.
      READ (BUF_IN, ID_NUM);               -- Read integer.
      READ (BUF_IN, DOLLARS);              -- Read real number.
      WRITE (BUF_OUT, DOLLARS);            -- Write real number.
      WRITE (BUF_OUT, ' ');                -- Write space character.
      WRITE (BUF_OUT, ID_NUM);             -- Write integer.
      WRITE (BUF_OUT, ' ');                -- Write space character.
      WRITE (BUF_OUT, LNAME);              -- Write string.
      WRITELINE (TO_FILE, BUF_OUT);        -- Write line to file.
    end loop;
    wait;
  end process;
end SIMPLE;
```

Function ENDFILE and procedures READLINE, READ, WRITELINE and WRITE are predefined in the package TEXTIO. The while loop goes through all the lines in the file, one at a time. The READLINE procedure reads one line into the buffer BUF_IN. The READ procedure reads each field in the buffer. The WRITE procedure writes out the values into the output buffer. The WRITELINE procedure writes the output buffer to the output file. Here is a test input to this program.

```
Soccer        4569889  5660.23
Basketball    1890942  3561.99
Hockey        8766567  8009.12
```

Here is the output produced in the output file.

```
5.660230e+0      4569889 Soccer
3.561990e+03     1890942 Basketball
8.009120e+03     8766567 Hockey
```

3.4 Operators

The predefined operators in the language are classified into the following six categories:

1. Logical operators
2. Relational operators
3. Shift operators
4. Adding operators

5. Multiplying operators
6. Miscellaneous operators

The operators have increasing precedence going from category 1 to 6. Operators in the same category have the same precedence, and evaluation is done left to right. Parentheses may be used to override the left to right evaluation.

3.4.1 Logical Operators

The seven logical operators are:

and **or** **nand** **nor** **xor** **xnor** **not**

These operators are defined for the predefined types BIT and BOOLEAN. They are also defined for one-dimensional arrays of BIT and BOOLEAN. During evaluation of logical operators, bit values '0' and '1' are treated as FALSE and TRUE values of the BOOLEAN type, respectively. The result of a logical operation has the same type as its operands. The **not** operator is a unary logical operator and has the same precedence as that of miscellaneous operators.

The operators **nand** and **nor** are not associative; therefore, the syntax of an expression with a sequence of **nand** or **nor** operators is illegal. For example, the following expression is illegal.

```
A nand B nand C                -- is illegal.
```

Parentheses can be used to avoid this problem.

3.4.2 Relational Operators

These are:

= /= < <= > >=

The result type for all relational operations is always the predefined type BOOLEAN. The = (equality) and the /= (inequality) operators are predefined on any type except file types. The remaining four relational operators are predefined on any scalar type (e.g., integer or enumerated types) or discrete array type (i.e., arrays in which element values belong to a discrete type). When operands are discrete array types, comparison is performed one element at a time from left to right. For example,

```
BIT_VECTOR'('0', '1', '1') < BIT_VECTOR'('1', '0', '1')
```

is true, since the first element in the first array aggregate is less than the first element in the second aggregate. Similarly, if

```
type MVL is ('U', '0', '1', 'Z');
```

then

```
MVL'('U') < MVL'('Z')
```

is true, since 'U' occurs to the left of 'Z'. Note that it is necessary to qualify the aggregates and literals used in these examples since the literals are overloaded and their type cannot be determined from its use. Qualified expressions are described in Chapter 10.

Here is an example in which operands are arrays of different lengths.

```
"VHDL" < "VHDL92"
```

is true. Comparison is again performed one element at a time from left to right. However, if a corresponding element does not exist in an operand, it is considered to be null, and null is always considered to be less than any character. In this example, the character '9' in the second operand has no corresponding element in the first operand.

3.4.3 Shift Operators

These are:

sll **srl** **sla** **sra** **rol** **ror**

Each of the operators takes an array of BIT or BOOLEAN as the left operand and an integer value as the right operand and performs the specified operation. If the integer value is a negative number, the opposite action is performed, that is, a left shift or rotate becomes a right shift or rotate, respectively, and vice versa.

The **sll** operator (shift left logical) and the **srl** operator (shift right logical) fill the vacated bits with *left-operand-type*'LEFT. The **sla** operator (shift left arithmetic) fills the vacated bits with the rightmost bit of the left operand, while the **sra** operator (shift right arithmetic) fills the vacated bits with the leftmost bit of the left operand. The rotate operators cause the vacated bits to be filled with the displaced bits in a circular fashion. Here are some examples.

```
-- Assume that all left operands are BIT_VECTORS:
"1001010" sll 2     is "0101000"        -- filled with BIT'LEFT, which is '0'.
```

```
"1001010" srl 3    is "0001001"    -- filled with '0'.
"1001010" sla 2    is "0101000"    -- filled with rightmost bit which is '0'.
"1001010" sra 3    is "1111001"    -- filled with '1' which is the leftmost bit.
"1001010" rol 2    is "0101010"    -- rotate left.
"1001010" ror 3    is "0101001"    -- rotate right.
"1001010" ror –4   is "0101001"    --"rol 4" operation performed.
"1001010" srl –5   is "1000000"    --"sll 5" operation performed.
"1001010" sla –2   is "1110010"    --"sra 2" operation performed.
"1001010" sra –4   is "0100000"    --"sla 4" operation performed.
"1001010" rol –1   is "0100101"    --"ror 1" operation performed.
"1001010" sll –5   is "0000010"    --"srl 5" operation performed.
```

3.4.4 Adding Operators

These are:

\+ – &

The operands for the + (addition) and – (subtraction) operators must be of the same numeric type, with the result being of the same numeric type. The addition and subtraction operators may also be used as unary operators, in which case the operand and the result type are the same. The operands for the & (concatenation) operator can be either a one-dimensional array type or an element type. The result is always an array type. For example,

```
'0' & '1'
```

results in an array of characters "01".

```
'C' & 'A' & 'T'
```

results in the value "CAT".

```
"BA" & "LL"
```

creates an array of characters "BALL".

A good example of an use of a concatenation operator is when many separate signals are to be grouped into one bus. Here is an example.

```
signal DBUS: STD_LOGIC_VECTOR (0 to 7);
signal CTRL: STD_LOGIC_VECTOR (1 downto 0);
signal ENABLE, RW: STD_LOGIC;
signal COUNT: STD_LOGIC_VECTOR (0 to 3);
-- This is the intention:
-- DBUS(0 to 1) <= CTRL;
-- DBUS(2) <= ENABLE;
```

```
-- DBUS(3) <= RW;
-- DBUS(4 to 7) <= COUNT;

-- Using the concatenation operator, we get:
DBUS <= CTRL & ENABLE & RW & COUNT;
```

3.4.5 Multiplying Operators

These are:

```
*     /     mod     rem
```

The * (multiplication) and / (division) operators are predefined for both operands being of the same integer or floating point type. The result is also of the same type. The multiplication operator is also defined for the case when one of the operands is of physical type and the second operand is of integer or real type. The result is of physical type.

For the division operator, division of a value of physical type by either an integer or a real value is allowed, and the result type is the physical type. Division of a value of physical type by another object of the same physical type is also defined, and it yields an integer value as a result.

The **rem** (remainder) and **mod** (modulus) operators operate on operands of integer types, and the result is also of the same type. The result of a **rem** operation has the sign of its first operand and is defined as:

```
A rem B = A – (A / B) * B
```

The result of a **mod** operator has the sign of the second operand and is defined as:

```
A mod B = A – B * N                -- For some integer N.
```

Here are some examples using the **mod** and **rem** operators.

```
7 mod 4            -- has value 3
(–7) rem 4         -- has value –3
7 mod (–4)         -- has value –1
(–7) rem (–4)      -- has value –3
```

3.4.6 Miscellaneous Operators

The miscellaneous operators are:

abs **

The **abs** (absolute) operator is defined for any numeric type.

The ** (exponentiation) operator is defined for the left operand to be of integer or floating point type, and for the right operand (i.e., the exponent) to be of integer type only.

The **not** logical operator has the same precedence as the above two operators.

❑

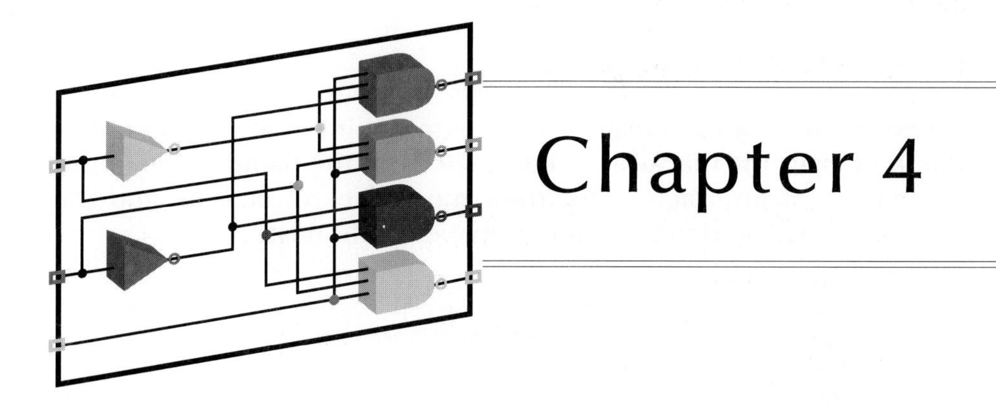

Chapter 4

Behavioral Modeling

This chapter presents the behavioral style of modeling. In this modeling style, the behavior of the entity is expressed using sequentially executed, procedural code, which is very similar in syntax and semantics to that of a high-level programming language like C or Pascal. A process statement is the primary mechanism used to model the behavior of an entity. This chapter describes the process statement and the various kinds of sequential statements that can be used within a process statement to model such behavior.

Irrespective of the modeling style used, every entity is represented using an entity declaration and at least one architecture body. The first two sections describe these in detail.

4.1 Entity Declaration

An entity declaration describes the external interface of the entity, that is, it gives the black-box view. It specifies the name of the entity, the names of interface ports, their mode (i.e., direction), and the type of ports. The syntax for an entity declaration is:

entity *entity-name* **is**
  [**generic** (*list-of-generics-and-their-types*) ;]
  [**port** (*list-of-interface-port-names-and-their-types*) ;]
  [*entity-item-declarations*]
[**begin**
  entity-statements]
end [**entity**] [*entity-name*] ;

Generics are discussed in Chapter 7, and *entity-statements* are discussed in Chapter 10. The *entity-name* is the name of the entity, and the interface ports are the signals through which the entity passes information to and from its external environment. Each interface port can have one of the following modes:

1. **in**: The value of an input port can only be read within the entity model.
2. **out**: The value of an output port can only be updated within the entity model; it cannot be read.
3. **inout**: The value of a bidirectional port can be read and updated within the entity model.
4. **buffer**: The value of a buffer port can be read and updated within the entity model. However, it differs from the **inout** mode in that it cannot have more than one source, and the only kind of signal that can be connected to it can be another buffer port or a signal with at most one source.
5. **linkage**: The value of a linkage port can be read and updated. This can be done only by another port of mode **linkage**. The usage of linkage ports is not very clear and is thus not recommended. In the past, it has been used to interface with foreign-language models and foreign simulators.

Declarations that are placed in the *entity-item-declarations* section are common to all the design units that are associated with that entity declaration (these may be architecture bodies and configuration declarations). Examples of these are given in Chapter 10. An And-Or-Invert circuit is shown in Figure 4-1, and its corresponding entity declaration is shown next.

```
entity AOI is
   port (A, B, C, D: in BIT; Z: out BIT);
end AOI;
```

The entity declaration specifies that the name of the entity is AOI, and that it has four input signals of type BIT and one output signal of type BIT. Note that it does not specify the composition or functionality of the entity.

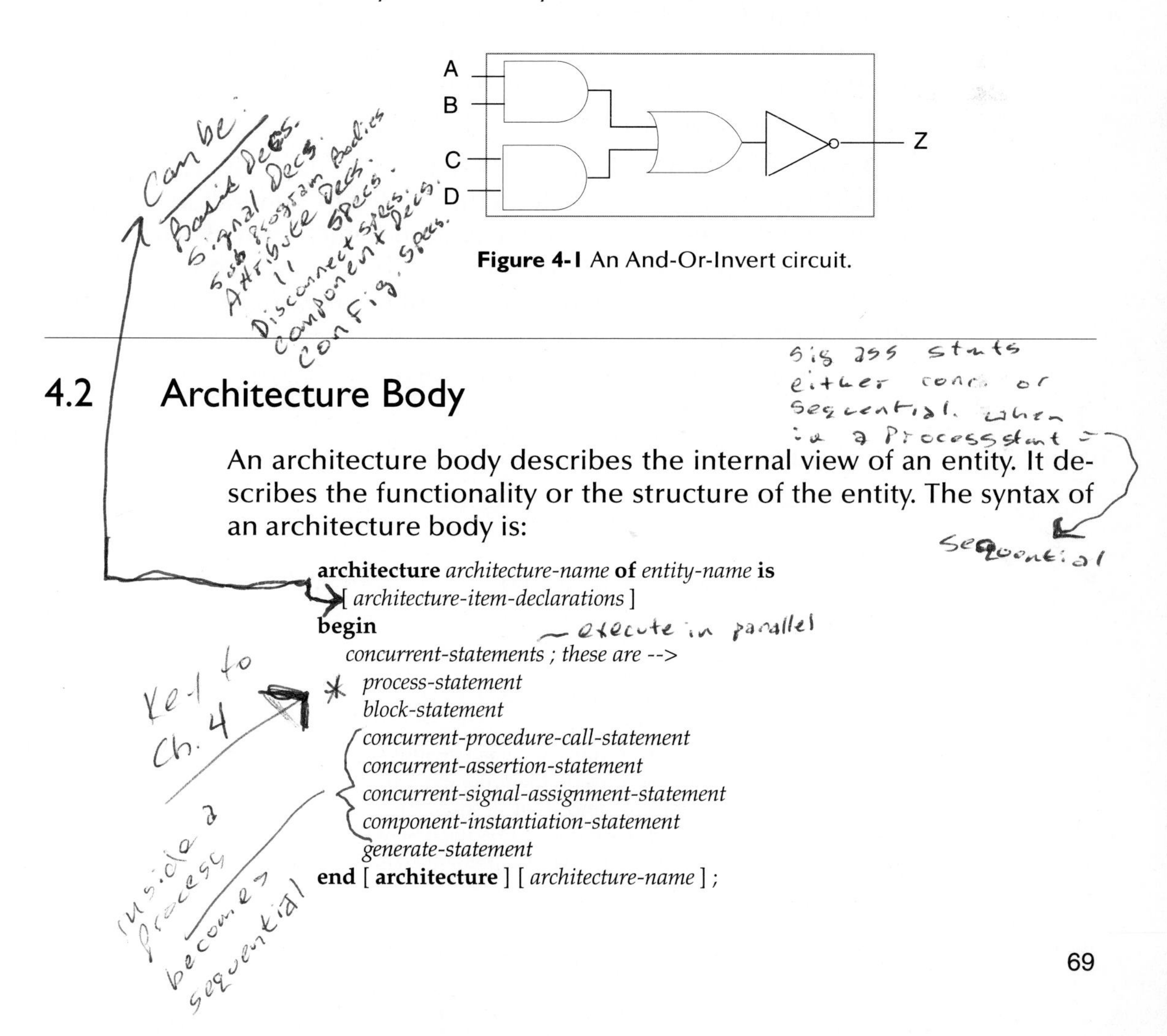

Figure 4-1 An And-Or-Invert circuit.

4.2 Architecture Body

An architecture body describes the internal view of an entity. It describes the functionality or the structure of the entity. The syntax of an architecture body is:

```
architecture architecture-name of entity-name is
   [ architecture-item-declarations ]
begin
   concurrent-statements ; these are -->
      process-statement
      block-statement
      concurrent-procedure-call-statement
      concurrent-assertion-statement
      concurrent-signal-assignment-statement
      component-instantiation-statement
      generate-statement
end [ architecture ] [ architecture-name ] ;
```

The concurrent statements describe the internal composition of the entity. All concurrent statements execute in parallel; therefore, their textual order of appearance within the architecture body has no impact on the implied behavior. The internal composition of an entity can be expressed in terms of structure, dataflow, and sequential behavior. These are described using concurrent statements. For example, component instantiations are used to express structure, concurrent signal assignment statements are used to express dataflow, and process statements are used to express sequential behavior. Each concurrent statement is a different element operating in parallel, similar to individual gates of a design operating in parallel.

The item declarations declare items that are available for use within the architecture body. The names of items declared in the entity declaration, including ports and generics, are available for use within the architecture body due to the association of the entity name with the architecture body by the statement:

```
architecture architecture-name of entity-name is . . .
```

An entity can have many internal views, each of which is described using a separate architecture body. In general, an entity is represented using one entity declaration (which provides the external view) and one or more architecture bodies (which provide the internal view). Here are two examples of architecture bodies for the same AOI entity.

```
architecture AOI_CONCURRENT of AOI is
begin
  Z <= not ( (A and B) or (C and D) );
end AOI_CONCURRENT;

architecture AOI_SEQUENTIAL of AOI is
begin
  process (A, B, C, D)
    variable TEMP1, TEMP2: BIT;
  begin
    TEMP1 := A and B;                  -- statement 1
    TEMP2 := C and D;                  -- statement 2
    TEMP1 := TEMP1 or TEMP2;           -- statement 3
    Z <= not TEMP1;                    -- statement 4
    -- It is possible to replace the above four statements with just
    -- one statement inside the process:
      -- Z <= not ((A and B) or (C and D));
    -- However, it has been used as such here to explain the
    -- sequential nature of the statements within a process.
```

```
    end process;
end AOI_SEQUENTIAL;
```

The first architecture body, AOI_CONCURRENT, describes the AOI entity using the dataflow style of modeling; the second architecture body, AOI_SEQUENTIAL, uses the behavioral style of modeling. In this chapter, we are concerned with describing an entity using the behavioral modeling style. A process statement, which is a concurrent statement, is the primary mechanism used to describe the functionality of an entity in this modeling style.

4.3 Process Statement

A process statement contains sequential statements that describe the functionality of a portion of an entity in sequential terms. The syntax of a process statement is:

```
[ process-label : ] process [ ( sensitivity-list ) ] [ is ]
    [ process-item-declarations ]
begin
    sequential-statements; these are -->
        variable-assignment-statement
        signal-assignment-statement
        wait-statement
        if-statement
        case-statement
        loop-statement
        null-statement
        exit-statement
        next-statement
        assertion-statement
        report-statement
        procedure-call-statement
        return-statement
end process [ process-label ] ;
```

A set of signals to which the process is sensitive is defined by the sensitivity list. In other words, each time an event occurs on any of the signals in the sensitivity list, the sequential statements within the process are executed in a sequential order, that is, in the order in which they appear (similar to statements in a high-level programming language like C or Pascal). The process then suspends after executing the last sequential statement and waits for another event to occur on a signal in the sensitivity list. Items declared in the item declarations part are available for use only within the process. Every sequential statement can optionally have a label.

The architecture body, AOI_SEQUENTIAL, presented earlier, contains one process statement. This process statement has four signals in its sensitivity list and one variable declaration. If an event occurs on any of the signals, A, B, C, or D, the process is executed. This is accomplished by executing statement 1 first, then statement 2, followed by statement 3, and then statement 4. After this, the process suspends (simulation does not stop, however) and waits for another event to occur on a signal in the sensitivity list.

4.4 Variable Assignment Statement

Variables can be declared and used inside a process statement. A variable is assigned a value using the variable assignment statement that typically has the form:

variable-object := *expression* ;

The expression is evaluated when the statement is executed, and the computed value is assigned to the variable object instantaneously, that is, at the current simulation time.

Variables are created at the time of elaboration and retain their values throughout the entire simulation run (like static variables in C). This is because a process is never exited; it is either in an active state, that is, being executed, or in a suspended state, that is, waiting for a certain event to occur. A process is first entered at the start of simulation (actually, during the initialization phase of simulation), at which time it is executed until it suspends because of a `wait` statement (`wait` statements are described later in this chapter) or a sensitivity list.

Consider the following process statement.

```
process (A)
  variable EVENTS_ON_A: INTEGER := –1;
begin
  EVENTS_ON_A := EVENTS_ON_A + 1;
end process;
```

At the start of simulation, the process is executed once. The variable EVENTS_ON_A gets initialized to –1 and then incremented by 1. After that, any time an event occurs on signal A, the process is activated and the single variable assignment statement is executed. This causes the variable EVENTS_ON_A to be incremented. At the end of simulation, variable EVENTS_ON_A contains the total number of events that occurred on signal A.

Here is another example of a process statement.

```
signal A, Z: INTEGER;
. . .
PZ: process (A)                  -- PZ is a label for the process.
   variable V1, V2: INTEGER;
begin
   V1 := A – V2;                 -- statement 1
   Z <= –V1;                     -- statement 2
   V2 := Z + V1 * 2;             -- statement 3
end process PZ;
```

If an event occurred on signal A at time T_1, and variable V2 was assigned a value, say 10, in statement 3, then the next time an event occurs on signal A, say, at time T_2, the value of V2 used in statement 1 would still be 10.

A variable can also be declared outside of a process or a subprogram. Such a variable can be read and updated by more than one process. These variables are called *shared variables*. Chapter 10 explains shared variables in more detail.

4.5 Signal Assignment Statement

Signals are assigned values using a signal assignment statement. The simplest form of a signal assignment statement is:

signal-object <= *expression* [**after** *delay-value*] ;

A signal assignment statement can appear within a process or outside of a process. If it occurs outside of a process, it is considered to be a concurrent signal assignment statement; this form is discussed in the next chapter. When a signal assignment statement appears within a process, it is considered to be a sequential signal assignment statement and is executed in sequence with respect to the other sequential statements which appear within that process.

When a signal assignment statement is executed, the value of the expression is computed, and this value is scheduled to be assigned to the signal after the specified delay. It is important to note that the expression is evaluated at the time the statement is executed (which is the current simulation time) and not after the specified delay. If no `after` clause ("**after** *delay-value*") is specified, the delay is assumed to be a default delta delay.

Some examples of signal assignment statements are:

```
COUNTER <= COUNTER + "0010";           -- Assign after a delta delay.
PAR <= PAR xor DIN after 12 ns;
Z <= (A0 and A1) or (B0 and B1) or (C0 and C1) after 6 ns;
```

Delta Delay

A *delta delay* is a very small delay (infinitesimally small). It does not correspond to any real delay, and actual simulation time does not advance. This delay models hardware where a minimal amount of time is needed for a change to occur, for example, in performing zero delay simulation. Delta delay allows for ordering of events that occur at the same simulation time during a simulation. Each unit of simulation time can be considered to be composed of an infinite number of delta delays. Therefore, an event always occurs at a real simulation time plus an integral multiple of delta delays. For example, events can occur at 15 ns, 15 ns + 1Δ, 15 ns + 2Δ, 15 ns + 3Δ, 22 ns, 22 ns + Δ, 27 ns, 27 ns + Δ, and so on.

Consider the AOI_SEQUENTIAL architecture body described in Section 4.2. Let us assume that an event occurs on input signal D (that is, there is a change of value on signal D) at simulation time *T*. Statement 1 is executed first, and TEMP1 is assigned a value immediately because it is a variable. Statement 2 is executed next, and TEMP2 is assigned a value immediately. Statement 3 is executed next; it uses the values of TEMP1 and TEMP2 computed in statements 1 and 2, respectively, to determine the new value for TEMP1. Finally, statement 4 is executed, which causes signal Z to get the value of its right-hand-side expression after a delta delay, that is, signal Z gets its value only at time $T+\Delta$. This is shown in Figure 4-2.

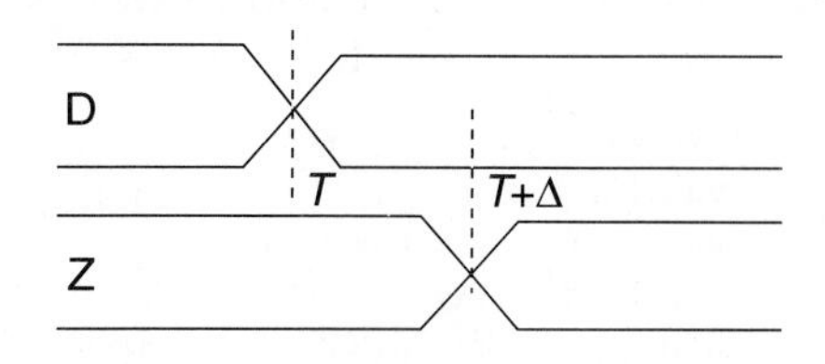

Figure 4-2 Delta delay.

Consider the process PZ described in the previous section. If an event occurs on signal A at time *T*, execution of statement 1 causes V1 to get a value, signal Z is then scheduled to get a value at time *T*+Δ, and finally, statement 3 is executed in which the old value of signal Z is used, that is, its value at time *T*, not the value that was scheduled to be assigned in statement 2. The reason for this is that simulation time is still at time *T* and has not advanced to time *T*+Δ. Later when simulation time advances to *T*+Δ, signal Z will get its new value. This example shows the important distinction between a variable assignment and a signal assignment statement. Variable assignments cause variables to get their values instantaneously, while signal assignments always cause signals to get their values at a later time (at least a delta delay later).

So far we have seen two examples of sequential statements, the variable assignment statement and the signal assignment statement. Other kinds of sequential statements are described next.

4.6 Wait Statement

As we saw earlier, a process may be suspended by means of a sensitivity list. That is, when a process has a sensitivity list, it always suspends after executing the last sequential statement in the process. The `wait` statement provides an alternate way to suspend the execution of a process. There are three basic forms of the `wait` statement.

wait on *sensitivity-list* ;
wait until *boolean-expression* ;
wait for *time-expression* ;

They may also be combined in a single `wait` statement. For example,

wait on *sensitivity-list* **until** *boolean-expression* **for** *time-expression* ;

Some examples of `wait` statements are:

```
wait on A, B, C;                        -- statement 1
wait until A = B;                       -- statement 2
wait for 10 ns;                         -- statement 3
wait on CLOCK for 20 ns;                -- statement 4
wait until SUM > 100 for 50 ms;         -- statement 5
wait on CLOCK until SUM > 100;          -- statement 6
```

The execution of the `wait` statement in statement 1 causes the enclosing process to suspend and then wait for an event to occur on signal A, B, or C. Once that happens, the process resumes execution from the next statement onwards. If the `wait` statement was the last statement in the process, the process resumes execution from the first statement.

The execution of the `wait` statement in statement 2 causes the enclosing process to suspend until the specified condition becomes true. When an event occurs on signal A or B, the condition is evaluated. If it is true, the process resumes execution from the next statement onwards, otherwise, it suspends again. When the `wait` statement in statement 3 is executed, say, at time *T*, the enclosing process suspends for 10 ns, and when simulation time advances to *T*+10 ns, the enclosing process resumes execution from the statement following the `wait` statement.

The execution of statement 4 causes the enclosing process to suspend and then wait for an event to occur on signal CLOCK for a time-out interval of 20 ns. If no event occurs within 20 ns, the process resumes execution with the statement following the wait. In statement 5, the process suspends for a maximum of 50 ms until the value of signal SUM is greater than 100. The Boolean condition is evaluated every time there is an event on signal SUM. If the Boolean condition is not satisfied for 50 ms, the process resumes from the statement following the `wait` statement.

In the last example of the `wait` statement (statement 6), it means to wait until an event occurs on CLOCK, then to check if SUM > 100. If SUM <= 100, continue to wait. If SUM > 100, resume from the statement following the `wait` statement.

It is possible for a process not to have an explicit sensitivity list. In such a case, the process may have one or more `wait` statements. It must have at least one `wait` statement; otherwise, the process will never get suspended and would remain in an infinite loop during the initialization phase of simulation. It is an error if both a sensitivity list and a `wait` statement are present within a process.

The presence of a sensitivity list in a process implies the presence of an implicit "**wait on** *sensitivity-list*" statement as the last statement in the process. An equivalent process statement for the process statement in the AOI_SEQUENTIAL architecture body is:

```
process                                    -- No sensitivity list.
  variable TEMP1, TEMP2: BIT;
begin
  TEMP1 := A and B;
  TEMP2 := C and D;
  TEMP1 := TEMP1 or TEMP2;
  Z <= not TEMP1;
  wait on A, B, C, D;                      -- Replaces the sensitivity list.
end process;
```

Therefore, a process with a sensitivity list always suspends at the end of the process and, when reactivated due to an event, resumes execution from the first statement in the process.

wait for 0

What does the following `wait` statement imply?

```
wait for 0 ns;
```

It means to wait for one delta cycle. This statement is useful when you want the process to be delayed so that delta-delayed signal assignments within a process can take effect. For example,

```
WAIT0: process
begin
  wait on DATA;
  SIG_A <= DATA;
  wait for 0 ns;
  SIG_B <= SIG_A;
end process;
```

If DATA changes at 10 ns, SIG_A is scheduled to get the new value of DATA at $10+1\Delta$. The `wait` statement ("**wait for** 0 ns") causes the process to suspend for one delta. SIG_A gets updated with it's new value. Process resumes at $10 + 1\Delta$. SIG_B gets scheduled to get the new value of SIG_A at $10+2\Delta$.

If the "**wait for** 0 ns" statement was not present, both the signal assignments are executed sequentially at time 10 ns, and SIG_B is scheduled to get the current value of SIG_A (which is not yet the new value of DATA) at time 10+1Δ. SIG_B always gets the value of SIG_A at the time DATA changes but after a delta delay. This effect is further elaborated in Figure 4-3.

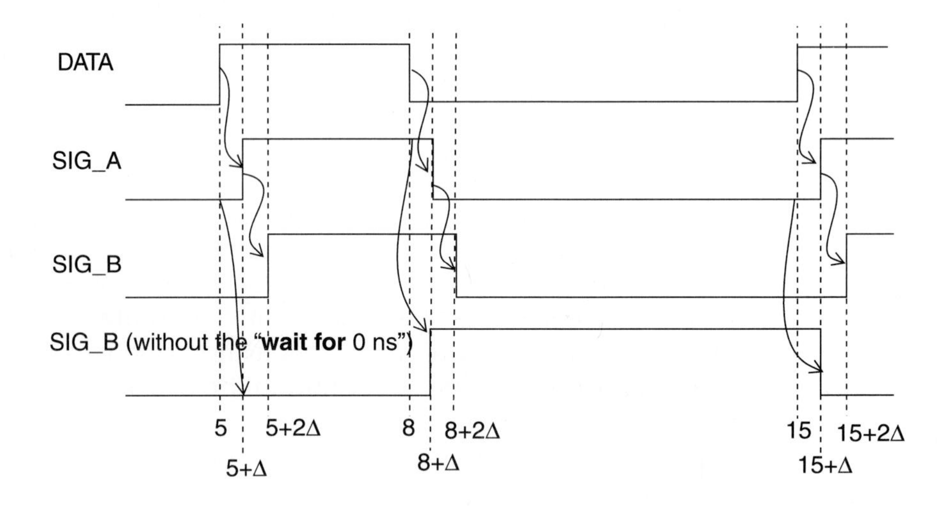

Figure 4-3 Effect of "**wait for** 0 ns".

wait until

The behavior of the "**wait until**" form is worth a second explanation. So let us look at an example.

wait until CLOCK = '1';

When the `wait` statement is executed, the process (enclosing) must first suspend, then wait for an event to occur on CLOCK, and then check the value of the boolean expression. It does not matter if the boolean expression is true at the time the `wait` statement is executed. So if CLOCK is already '1' when the `wait` statement is executed, the boolean expression is not checked until another event occurs on signal CLOCK.

For the same reason,

wait until TRUE;

when executed will suspend the process indefinitely (since no signal is present in the boolean expression and hence no event can occur).

4.7 If Statement

An `if` statement selects a sequence of statements for execution based on the value of a condition. The condition can be any expression that evaluates to a Boolean value. The general form of an `if` statement is:

```
if boolean-expression then
   sequential-statements
{ elsif boolean-expression then          -- elsif clause; if statement can
   sequential-statements }                -- have 0 or more elsif clauses.
[ else                                    -- else clause (optional).
   sequential-statements ]
end if;
```

The `if` statement is executed by checking each condition sequentially until the first true condition is found; then, the set of sequential statements associated with this condition is executed. The `if` statement can have zero or more `elsif` clauses and an optional `else` clause. An `if` statement is also a sequential statement, and therefore, the previous syntax allows for arbitrary nesting of `if` statements. Here are some examples.

```
if SUM <= 100 then                 --"<="is a less-than-or-equal-to operator.
   SUM := SUM + 10;
end if;

if NICKEL_IN then
   DEPOSITED <= TOTAL_10;          -- This"<=" is a signal assignment
                                   -- operator.
elsif DIME_IN then
   DEPOSITED <= TOTAL_15;
elsif QUARTER_IN then
   DEPOSITED <= TOTAL_30;
else
   DEPOSITED <= TOTAL_ERROR;
end if;

if CTRL1 = '1' then
   if CTRL2 = '0' then
      MUX_OUT <= "0010";
```

```
    else
      MUX_OUT <= "0001";
    end if;
  else
    if CTRL2 = '0' then
      MUX_OUT <= "1000";
    else
      MUX_OUT <= "0100";
    end if;
  end if;
```

A complete example of a two-input `nor` gate entity using an `if` statement is shown next.

```
entity NOR2 is
  port (A, B: in BIT; Z: out BIT);
end NOR2;

architecture NOR2 of NOR2 is               -- Architecture body can have
                                           -- same name as entity.
begin
  P1: process (A, B)
    constant RISE_TIME: TIME := 10 ns;
    constant FALL_TIME: TIME := 8 ns;
  begin
    if (A nor B) = '1' then
      Z <= '1' after RISE_TIME;
    else
      Z <= '0' after FALL_TIME;
    end if;
  end process P1;
end NOR2;
```

4.8 Case Statement

The format of a `case` statement is:

```
case expression is
  when choices => sequential-statements       -- branch #1
  when choices => sequential-statements       -- branch #2
  -- Can have any number of branches.
  [ when others => sequential-statements ]    -- last branch
end case;
```

The `case` statement selects one of the branches for execution based on the value of the expression. The expression value must be of a discrete type or of a one-dimensional array type. Choices may be expressed as single values, as a range of values, by using | (vertical bar: represents an "or" choice), or by using the others clause ("**when others**"). All possible values of the expression must be covered in the case statement exactly once. The others clause can be used as a choice to cover the "catch-all" values and, if present, must be the last branch in the case statement. An example of a case statement is:

```
type WEEK_DAY is (MON, TUE, WED, THU, FRI, SAT, SUN);
type DOLLARS is range 0 to 10;
variable DAY: WEEK_DAY;
variable POCKET_MONEY: DOLLARS;

case DAY is
   when TUE          => POCKET_MONEY := 6;        -- branch 1
   when MON | WED    => POCKET_MONEY := 2;        -- branch 2
   when FRI to SUN   => POCKET_MONEY := 7;        -- branch 3
   when others       => POCKET_MONEY := 0;        -- branch 4
end case;
```

Branch 2 is chosen if DAY has the value of either MON or WED. Branch 3 covers the values FRI, SAT, and SUN, while branch 4 covers the remaining value, THU. The case statement is also a sequential statement, and it is therefore possible to have nested case statements. A model for a 4-by-1 multiplexer using a case statement is shown next.

```
library IEEE;
use IEEE.STD_LOGIC_1164.all;
entity MUX is
   port ( A, B, C, D     : in STD_LOGIC;
          CTRL           : in STD_LOGIC_VECTOR(0 to 1);
          Z              : out STD_LOGIC);
end MUX;

architecture MUX_BEHAVIOR of MUX is
   constant MUX_DELAY: TIME := 10 ns;
begin
   PMUX: process (A, B, C, D, CTRL)
      variable TEMP: STD_LOGIC;
   begin
      case CTRL is
         when "00" => TEMP := A;
         when "01" => TEMP := B;
         when "10" => TEMP := C;
         when "11" => TEMP := D;
```

```
          when others => TEMP := 'X';         -- Covers values such as
                                              -- "0Z", "ZU", "U1", etc.
        end case;
        Z <= TEMP after MUX_DELAY;
      end process PMUX;
    end MUX_BEHAVIOR;
```

4.9 Null Statement

The statement

```
null ;
```

is a sequential statement that does not cause any action to take place; execution continues with the next statement. One example of this statement's use is in an `if` statement or in a `case` statement where, for certain conditions, it may be useful or necessary to explicitly specify that no action needs to be performed.

4.10 Loop Statement

A `loop` statement is used to iterate through a set of sequential statements. The syntax of a `loop` statement is:

```
[ loop-label : ] iteration-scheme loop
   sequential-statements
end loop [ loop-label ] ;
```

There are three types of iteration schemes. The first is the `for` iteration scheme, which has the form:

```
for identifier in range
```

An example of this iteration scheme is:

```
FACTORIAL := 1;
for NUMBER in 2 to N loop
   FACTORIAL := FACTORIAL * NUMBER;
end loop;
```

In this example, the body of the `for` loop is executed (N–1) times, with the loop identifier, NUMBER, being incremented by 1 at the end of each iteration. The object NUMBER is implicitly declared within the `for` loop to belong to the integer type whose values are in the range 2 to *N*. Therefore, no explicit declaration for the loop identifier is necessary. The loop identifier also cannot be assigned any value inside the `for` loop. If another variable with the same name exists outside the `for` loop, these two variables are treated separately, and the variable used inside the `for` loop refers to the loop identifier.

The range in a `for` loop can also be a range of an enumeration type, such as the one shown in the following example.

```
type HEXA is ('0', '1', '2', '3', '4', '5', '6', '7', '8', '9', 'A', 'B', 'C', 'D', 'E', 'F');
...
for NUM in HEXA'('9') downto HEXA'('0') loop
   -- NUM will take values in type HEXA from '9' through '0'.
   ...
end loop;

for CHAR in HEXA loop
   -- CHAR will take all values in type HEXA from '0' through 'F'.
   ...
end loop;
```

Notice that it is necessary to qualify the values being used for NUM (e.g., HEXA'('9')) since the literals '0' through '9' are overloaded, being defined once in type HEXA and the second time in the predefined type CHARACTER. Qualified expressions are described in Chapter 10.

The second form of the iteration scheme is the `while` scheme, which has the form:

```
while boolean-expression
```

An example of the `while` iteration scheme is:

```
J := 0; SUM := 10;
WH_LOOP: while J < 20 loop          -- This loop has a label, WH_LOOP.
   SUM := SUM * 2;
   J := J + 3;
end loop;
```

The statements within the body of the loop are executed sequentially and repeatedly as long as the loop condition, J < 20, is true. If the loop condition becomes false, execution continues with the statement following the `loop` statement.

The third and final form of the iteration scheme is one where no iteration scheme is specified. In this form of `loop` statement, all statements in the loop body are repeatedly executed until some other action causes the loop to exit. This action can be caused by an `exit` statement, a `next` statement, or a `return` statement. Here is an example.

```
SUM := 1; J := 0;
L2: loop                  -- This loop also has a label.
   J := J + 21;
   SUM := SUM * 10;
   exit when SUM > 100;
end loop L2;              -- This loop label, if present, must be the same
                          -- as the initial loop label.
```

In this example, the `exit` statement causes the execution to jump out of loop L2 when SUM becomes greater than 100. If the `exit` statement were not present, the loop would execute indefinitely.

4.11 Exit Statement

The `exit` statement is a sequential statement that can be used only inside a loop. It causes execution to jump out of the innermost loop or the loop whose label is specified. The syntax for an `exit` statement is:

exit [*loop-label*] [**when** *condition*] ;

If no loop label is specified, the innermost loop is exited. If the `when` clause ("**when** *condition*") is used, the specified loop is exited only if the given condition is true; otherwise, execution continues with the next statement. An alternate form for loop L2 described in the previous section is:

```
SUM := 1; J := 0;
L3: loop
   J : = J + 21;
   SUM := SUM * 10;

   if SUM > 100 then
      exit L3;                        --"exit;" is also sufficient.
   end if;
end loop L3;
```

Here is another example of an `exit` statement.

```
loop
   wait on A, B;
   exit when A = B;          -- Since no label, innermost loop is exited.
end loop;
-- This loop behaves exactly like the wait statement:
-- wait until A = B;
```

4.12 Next Statement

The `next` statement is also a sequential statement that can be used only inside a loop. The syntax is the same as that for the `exit` statement except that the keyword **next** replaces the keyword **exit**. Its syntax is:

> **next** [*loop-label*] [**when** *condition*] ;

The `next` statement results in skipping the remaining statements in the current iteration of the specified loop; execution resumes with the first statement in the next iteration of this loop, if one exists. If no loop label is specified, the innermost loop is assumed. In contrast to the `exit` statement, which causes the loop to be terminated (i.e., exits the specified loop), the `next` statement causes the current loop iteration of the specified loop to be prematurely terminated; execution resumes with the next iteration. Here is an example.

```
for J in 10 downto 5 loop
   if SUM < TOTAL_SUM then
      SUM := SUM + 2;
   elsif SUM = TOTAL_SUM then
      next;
   else
      null;
   end if;
   K := K + 1;
end loop;
```

When the `next` statement is executed, execution jumps to the end of the loop (the last statement, K := K + 1, is not executed), decrements the value of the loop identifier, J, and resumes loop execution with this new value of J.

The next statement can also cause an inner loop to be exited. Here is such an example.

```
L4: for K in 10 downto 1 loop
  -- statements section 1
  L5: loop
    -- statements section 2
    next L4 when WR_DONE = '1';
    -- statements section 3
  end loop L5;
  -- statements section 4
end loop L4;
```

When WR_DONE = '1' becomes true, statements sections 3 and 4 are skipped, and execution jumps to the beginning of the next iteration of loop L4. Notice that the loop L5 was terminated because of the result of the next statement.

4.13 Assertion Statement

Assertion statements are useful in modeling constraints of an entity. For example, you may want to check if a signal value lies within a specified range, or check the setup and hold times for signals arriving at the inputs of an entity. If the check fails, a message is reported. The syntax of an assertion statement is:

```
assert boolean-expression
  [ report string-expression ]
  [ severity expression ] ;
```

If the value of the Boolean expression is false, the report message is printed along with the severity level. The expression in the severity clause must be a value of type SEVERITY_LEVEL (a predefined enumeration type with values NOTE, WARNING, ERROR, and FAILURE). The severity level is typically used by a simulator to initiate appropriate actions depending on its value. For example, if the severity level is ERROR, the simulator may abort the simulation process and provide relevant diagnostic information. At the very least, the severity level is displayed.

Here is a model of a D-type rising-edge-triggered flip-flop that uses assertion statements to check for setup and hold times.

```
entity DFF is
  port (D, CK: in BIT; Q, NOTQ: out BIT);
end DFF;

architecture CHECK_TIMES of DFF is
  constant HOLD_TIME: TIME := 5 ns;
  constant SETUP_TIME: TIME := 3 ns;
begin
  process (D, CK)
    variable LastEventOnD, LastEventOnCk: TIME;
  begin
    -- Check for hold time:
    if D'EVENT then
      assert NOW = 0 ns or
            (NOW – LastEventOnCk) >= HOLD_TIME
        report "Hold time too short!"
        severity FAILURE;
      LastEventOnD := NOW;
    end if;

    -- Check for setup time:
    if CK = '1' and CK'EVENT then
      assert NOW = 0 ns or
            (NOW – LastEventOnD) >= SETUP_TIME
        report "Setup time too short!"
        severity FAILURE;
      LastEventOnCk := NOW;
    end if;

    -- Behavior of FF:
    if CK = '1' and CK'EVENT then
      Q <= D;
      NOTQ <= not D;
    end if;
  end process;
end CHECK_TIMES;
```

'EVENT is a predefined attribute of a signal and is true if an event (a change of value) occurred on that signal at the time the value of the attribute is determined. Attributes are described in greater detail in Chapter 10. NOW is a predefined function that returns the current simulation time. In the previous example, the process is sensitive to signals D and CK. When an event occurs on either of these signals, the first `if` statement is executed. This checks to see if an event occurred on D. If so, the assertion is checked to make sure that the difference between the current simulation time and the last time an event occurred on signal CK is greater than a constant HOLD_TIME delay. If not, a report message is printed and the severity level is returned to the simulator. Similarly, the next `if` statement checks for the setup time. The last `if` statement describes the latch behavior of the D-type flip-flop. The setup and hold times have been modeled as constants in this example. These could also be modeled as generic parameters of the flip-flop. Generics are discussed in Chapter 7. The expression "(NOW = 0 ns)" is used in the assertion statements to prevent misleading messages appearing during the initialization phase of simulation (remember that all processes are executed until they suspend during the initialization phase).

Here is another example that uses an assertion statement to check for spikes at the input of an inverter.

```
package PACK1 is
  constant MIN_PULSE: TIME := 5 ns;
  constant PROPAGATE_DLY: TIME := 10 ns;
end PACK1;

use WORK.PACK1.all;
entity INV is
  port (A: in BIT; NOT_A: out BIT);
end INV;

architecture CHECK_INV of INV is
begin
  process (A)
    variable LastEventOnA: TIME := 0 ns;
  begin
    assert NOW = 0 ns or
          (NOW – LastEventOnA) >= MIN_PULSE
      report "Spike detected on input of inverter"
      severity WARNING;
    LastEventOnA := NOW;
    NOT_A <= not A after PROPAGATE_DLY;
  end process;
end CHECK_INV;
```

Some other examples of assertion statements are:

```
assert DATA <= 255
  report "Data out of range.";

assert (CLK = '0') or (CLK = '1');          -- CLK is of type STD_ULOGIC.
```

In the last assertion statement example, the default report message "Assertion violation" is printed. The default severity level is ERROR if the severity clause is not specified as in the previous two examples.

4.14 Report Statement

A report statement can be used to display a message. It is similar to an assertion statement, but without the assertion check. The syntax is of the form:

```
report string-expression
  [ severity expression ] ;
```

The expression in the severity clause, if present, must be of the predefined type SEVERITY_LEVEL. If not present, the default severity level of NOTE is used.

When a report statement is executed, it causes the specified string to be printed and the severity level to be reported to the simulator for appropriate action. Here are some examples.

```
if CLR = 'Z' then
  report "Signal CLR has a high-impedance value.";
  -- Default severity level is NOTE.
end if;

if CLK /= '0' and CLK /= '1' then
  report "CLK is neither a '0' nor a '1'!!!!"
    severity ERROR;
end if;
```

4.15 More on Signal Assignment Statement

Earlier in this chapter we saw a simple example of a signal assignment statement used inside a process that illustrated the delta delay model. In this section, we explain the signal assignment statement in greater detail to cover its more complex features. Aside from the delta delay model, there are two other types of delay models that can be used with signal assignments, inertial and transport.

4.15.1 Inertial Delay Model

Inertial delay models the delays often found in switching circuits. An input value must be stable for a specified *pulse rejection limit* duration before the value is allowed to propagate to the output. In addition, the value appears at the output after the specified inertial delay. If the input is not stable for the specified limit, no output change occurs. When used with signal assignments, the input value is represented by the value of the expression on the right-hand-side and the output is represented by the target signal. The syntax of such a specification in a signal assignment is:

signal-object <= [[**reject** *pulse-rejection-limit*] **inertial**] *expression*
after *inertial-delay-value* ;

If no pulse rejection limit is specified, the default pulse rejection limit is the inertial delay value itself. The pulse rejection limit cannot be negative or greater than the value of the inertial delay.

Figure 4-4 shows a simple example of a non-inverting buffer with an inertial delay of 10 ns and a pulse rejection limit of 4 ns.

```
Z <= reject 4 ns inertial A after 10 ns;
```

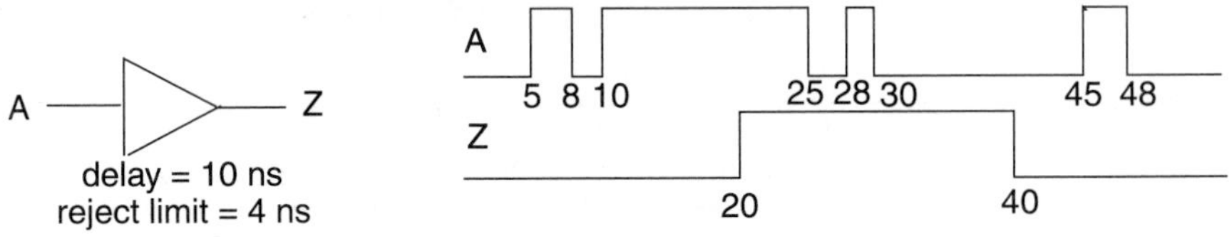

A Z
delay = 10 ns
reject limit = 4 ns

A
5 8 10 25 28 30 45 48
Z
20 40

Figure 4-4 Inertial delay example.

Events on signal A that occur at 5 ns and 8 ns are not stable for the duration of the pulse rejection limit, and hence do not propagate to the output. The event on A at 10 ns remains stable for more than the pulse rejection limit; therefore, the value is propagated to the target signal Z after the inertial delay - Z gets the value '1' at 20 ns. Events on signal A at 25 ns and 28 ns do not affect the output, since they are not stable for the duration of the pulse rejection limit. The '1' to '0' transition at time 30 ns on signal A remains stable for at least the duration of the pulse rejection limit, and therefore a '0' is propagated to signal Z with a delay of 10 ns; Z gets the new value at 40 ns. Other events on A do not affect the target signal Z. In this example, if the assignment had no reject clause specified, such as

```
Z <= A after 10 ns;
```

then the pulse rejection limit is also 10 ns, which is the inertial delay value.

Since inertial delay is most commonly found in digital circuits, it is the default delay model; that is, no keyword need be explicitly specified. It is possible, however, to explicitly identify the delay in a signal assignment as an inertial delay, even though it is not necessary, by using the keyword **inertial** explicitly, such as:

```
Z <= inertial A after 10 ns;
-- Same as: Z <= A after 10 ns;
```

This delay model is often used to filter out unwanted spikes and transients on signals.

4.15.2 Transport Delay Model

Transport delay models the delays in hardware that do not exhibit any inertial delay. This delay represents pure propagation delay; that is, any changes on an input are transported to the output, no matter how small, after the specified delay. To use a transport delay model, the keyword **transport** must be used in a signal assignment statement. Figure 4-5 shows an example of a non-inverting buffer using a transport delay of 10 ns.

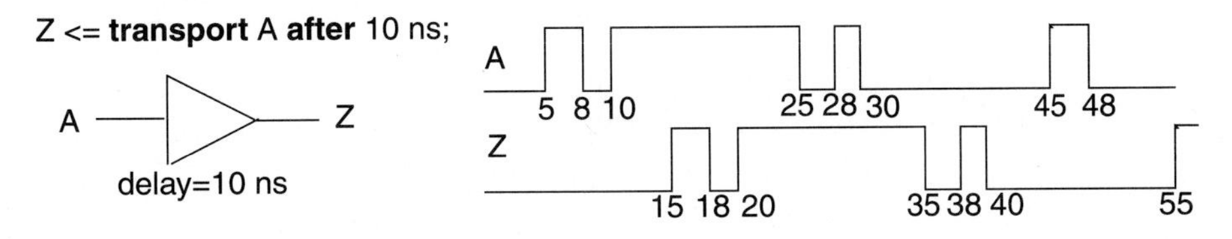

Figure 4-5 Transport delay example.

Ideal delay modeling can be obtained by using this delay model. In this case, spikes would be propagated through instead of being ignored, as in the inertial delay case. Routing delays can be modeled using transport delay. An example of a routing delay model is:

```
entity WIRE14 is
   port (A: in BIT; Z: out BIT);
end WIRE14;

architecture WIRE14_TRANSPORT of WIRE14 is
begin
   process (A)
   begin
      Z <= transport A after 0.1 ms;
   end process;
end WIRE14_TRANSPORT;
```

4.15.3 Creating Signal Waveforms

In all the examples of signal assignment statements that we have seen so far, we have always assigned a single value to a signal; this need not be so. It is possible to assign multiple values to a signal, each with a different delay value. For example,

```
PHASE1 <= '0', '1' after 8 ns, '0' after 13 ns, '1' after 50 ns;
```

When this signal assignment statement is executed, say, at time *T*, it causes four values to be scheduled for signal PHASE1; the value '0' is scheduled to be assigned at time $T+\Delta$, '1' at T+8 ns, '0' at T+13 ns, and '1' at T+50 ns. Thus, a waveform appears on the signal PHASE1, as shown in Figure 4-6.

A more general syntax of a signal assignment statement is:

```
signal-object <= [ transport | [ reject pulse-rejection-limit ] inertial ]
                 waveform-element , waveform-element ,
                 waveform-element , waveform-element ;
   -- Can have any number of waveform elements.
```

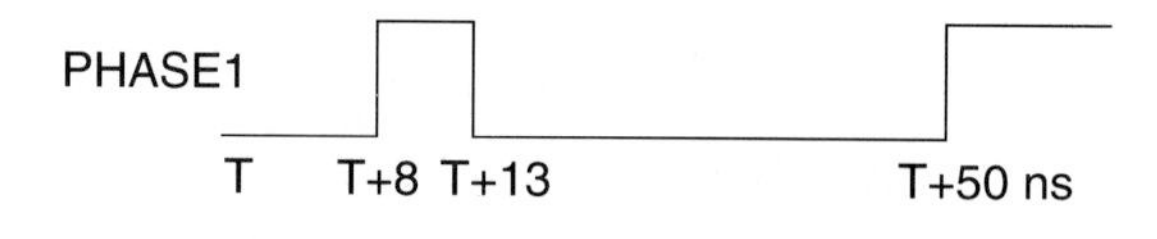

Figure 4-6 A signal waveform.

Each waveform element has a value part, specified by an expression (called the waveform expression in this text), and a delay part, specified by an `after` clause that specifies the delay. The delays in the waveform elements must appear in increasing order. A waveform element is of the form:

```
expression [ after time-expression ]
```

If no pulse rejection limit is specified in an assignment using inertial delays, the delay of the first waveform element becomes the default pulse rejection limit.

Any arbitrary waveform can therefore easily be created using a signal assignment statement.

4.15.4 Signal Drivers

What if there is more than one assignment to the same signal within a process? To understand how the waveform elements contribute to the effective value of a signal, it is important to understand the concept of a driver. A *driver* is created for every signal that is assigned a value in a process. The driver of a signal holds its current value and all its future values as a sequence of one or more transactions, where each transaction identifies the value to appear on the signal along with the time at which the value is to appear.

Consider the following signal assignment statement with three waveform elements and the driver which is created for that signal.

```
process
begin
  . . .
  RESET <= 3 after 5 ns, 21 after 10 ns, 14 after 17 ns;
end process;
```

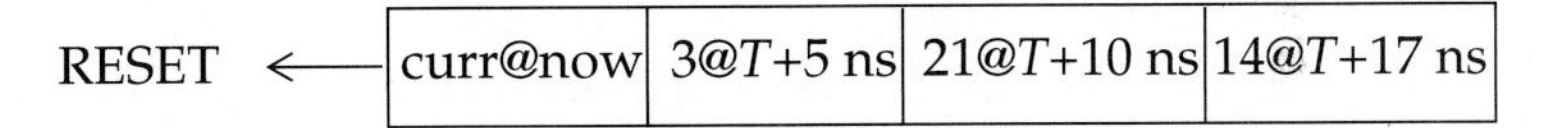

All transactions on the driver are ordered in increasing order of time. A driver always contains at least one transaction, which could be the initial value of the signal. The value of a signal is the value of its driver (the case where a signal has more than one driver is considered in the next chapter). In the previous example, when the signal assignment statement is executed, say, at time *T*, three new transactions are added to the driver for the RESET signal. The first transaction is the current value of the signal.

When simulation time advances to *T*+5 ns, the first transaction is deleted from the driver, and RESET gets the value of 3. When time advances to *T*+10 ns, the second transaction is deleted, and RESET gets the value of 21. When time advances to *T*+17 ns, the third transaction is deleted, and RESET gets the value of 14.

What if there is another signal assignment to RESET in the same process? Since a signal inside a process has only one driver, the transactions of the second signal assignment modify the transactions already present on the driver, depending on whether inertial or transport delay model is used. We consider the transport delay case first.

Effect of Transport Delay on Signal Drivers

Here is an example of a process having three signal assignments to the same signal RX_DATA.

```
signal RX_DATA: NATURAL;
. . .
process
begin
  . . .
  RX_DATA <= transport 11 after 10 ns;
  RX_DATA <= transport 20 after 22 ns;
  RX_DATA <= transport 35 after 18 ns;
end process;
```

Assume that the statements are executed at time *T*. The transactions on the driver for RX_DATA are created as follows. When the first signal assignment is executed, the transaction 11@*T*+10 ns is added to the driver. After the second signal assignment is executed, the transaction 20@*T*+ 22 ns is appended to the driver since the delay of this transaction (= 22 ns) is larger than the delay of the pending transactions on the driver. The driver for RX_DATA looks like this:

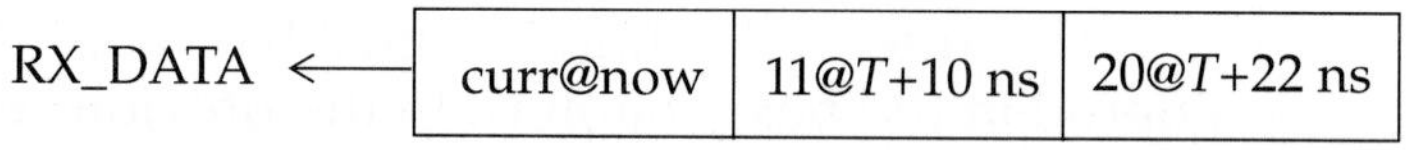

RX_DATA ⟵

curr@now	11@*T*+10 ns	20@*T*+22 ns

When the third signal assignment statement is executed, the new transaction 35@*T*+18 ns causes the 20@*T*+22 ns transaction to be deleted, and the new transaction is appended to the driver. The reason for this is that the delay for the new transaction (=18 ns) is less than the delay of the last transaction sitting on the driver (=22 ns). This effect is caused because transport delay is used. In general, a new transaction will delete all transactions sitting on a driver that are to occur at or later than the delay of the new transaction. Therefore, the driver for RX_DATA is changed to:

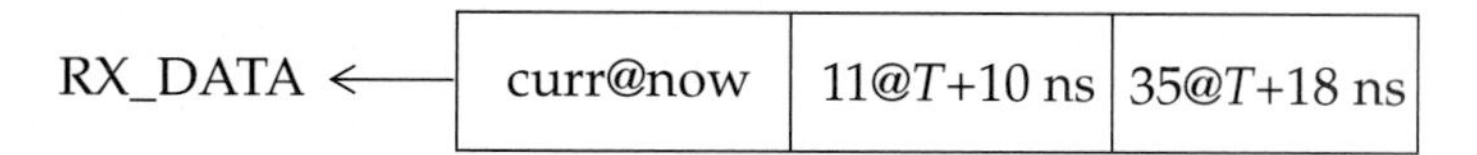

Here is another example.

```
process
   -- DATA_BUS is a signal of type BIT_VECTOR.
begin
   . . .
   DATA_BUS <= transport X"01" after 5 ns, X"FA" after 10 ns,
                     X"E8" after 15 ns;
   DATA_BUS <= transport X"B5" after 12 ns;
end process;
```

When these statements are executed, the status of the driver for DATA_BUS after the first signal assignment is:

DATA_BUS ← | curr@now | X"01"@*T*+5 ns | X"FA"@*T*+10 ns | X"E8"@*T*+15 ns |

When the second signal assignment is executed, the new transaction X"B5" @*T*+ 12 ns causes the X"E8" @*T*+15 ns transaction to be deleted from the driver (because its delay value is larger than the delay of the new transaction). The remaining transactions on the driver are not deleted, since their delays are less than the delay of the new transaction. Therefore, the final state of the driver after both the statements are executed is:

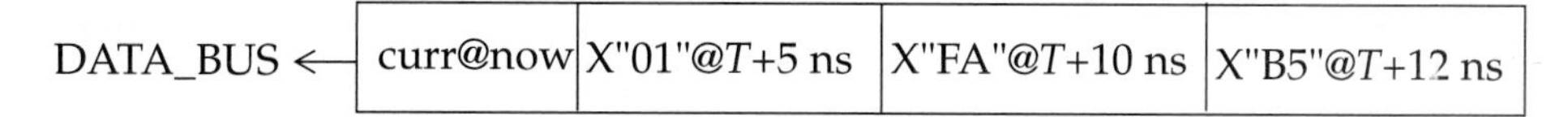

The following summarizes the rules for adding transactions from a signal assignment, using the transport delay mode, to a driver:

1. All transactions on the driver that occur at or after the delay time of the first new transaction are deleted.
2. All the new transactions are added at the end of the driver.

Effect of Inertial Delay on Signal Drivers

When inertial delays are used, both the signal value being assigned and the delay value affect the deletion and addition of transactions. If the delay of the first new transaction is earlier than an existing transaction, the latter is deleted regardless of the signal values of the two transactions. Note that this is the same as rule 1 for the transport delay case. On the other hand, existing transactions that fall within the window of the time of the first new transaction, say, F and (F – pulse rejection limit), are checked for their values. If their values are different, these already existing transactions are deleted; otherwise, they are retained. The following examples will help clarify this.

Consider the following process statement.

```
process
begin
   TX_DATA <= 11 after 10 ns;                 -- Pulse rejection limit is 10 ns.
   TX_DATA <= reject 15 ns inertial 22 after 20 ns;
   TX_DATA <= 33 after 15 ns;                 -- Pulse rejection limit is 15 ns.
   wait;                                      -- Suspends indefinitely.
end process;
```

The transaction 11@10 ns first gets added to the driver. The second transaction, 22@20 ns, causes the 11@10 ns transaction on the driver to be deleted. This is because the only transaction that falls in the window of 20 ns (the time of the first new transaction) and (20 ns – 15 ns) (the time of the first new transaction minus the pulse rejection limit) is the 11@10 ns transaction, and its value (which is 11) is different from the value of the first new transaction being added (which is 22). The state of the driver at this point is:

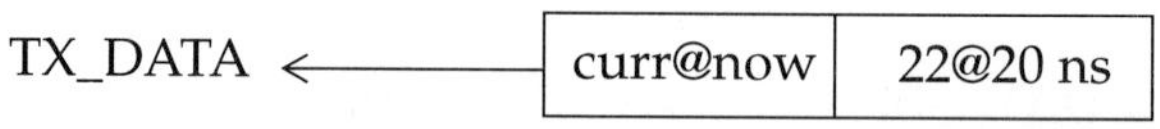

```
TX_DATA <------| curr@now | 22@20 ns |
```

The execution of the third signal assignment causes the transaction 22@20 ns to be deleted from the driver, since the delay of the new transaction (=15 ns) is less than the delay of the transaction on the driver (similar to the transport delay case). The final status of the driver is:

TX_DATA ←	curr@now	33@15 ns

Let us consider another example.

```
process
begin
  ADDR_BUS <= 1 after 5 ns, 21 after 9 ns, 6 after 10 ns,
        12 after 19 ns;        -- Pulse rejection limit is 5 ns, the
                               -- delay of the first waveform element.
  ADDR_BUS <= reject 4 ns inertial 6 after 12 ns, 20 after 19 ns;
  wait;
end process;
```

The status of the driver for the ADDR_BUS signal after the first statement is executed is:

ADDR_BUS ←	curr@now	1@5 ns	21@9 ns	6@10 ns	12@19 ns

The execution of the second statement causes all transactions on the driver later than 12 ns (the inertial delay of the first waveform element) to be deleted, which would be the 12@19 ns transaction. The two new transactions, 6@12 ns and 20@19 ns, are added to the driver. The old transaction, 21@9 ns, also gets deleted since its delay falls in the window of time 12 ns and (12 ns – 4 ns), and its value (which is 21) is not the same as the value of the first new transaction (which is 6).

The resulting driver is:

ADDR_BUS ←	curr@now	1@5 ns	6@10 ns	6@12 ns	20@19 ns

The summary of rules for adding transactions from a signal assignment using inertial delays are the following:

1. All transactions on a driver that are scheduled to occur at or after the delay of the first new transaction are deleted (as in the transport case).
2. Add all the new transactions to the driver.

3. For all the old transactions on the driver that occur at times between the time of the first new transaction (say F) and F minus the pulse rejection limit, delete the old transactions whose value is different from the value of the first new transaction.

4.16 Other Sequential Statements

There are two other forms of sequential statements:

1. Procedure call statement
2. Return statement

These are discussed in Chapter 8.

4.17 Multiple Processes

Since a process statement is a concurrent statement, it is possible to have more than one process within an architecture body. This makes it possible to capture the behavior of interacting processes, for example, that of interacting finite-state machines. Processes within an architecture body communicate with each other using signals that are visible to all the processes. Variables declared within a process cannot be used to pass information between processes, because their scope is limited to be within a process. Shared variables, however, could be used; these are described in Chapter 10.

Consider the following example of two interacting processes: RX, a receiver, and MP, a microprocessor. The RX process reads the serial input data and sends a READY signal indicating that the data has been read into the MP process. The MP process, after it assigns the data to the output, sends an acknowledge signal, ACK, back to the RX process to begin reading new input data. The block diagram for the two processes is shown in Figure 4-7.

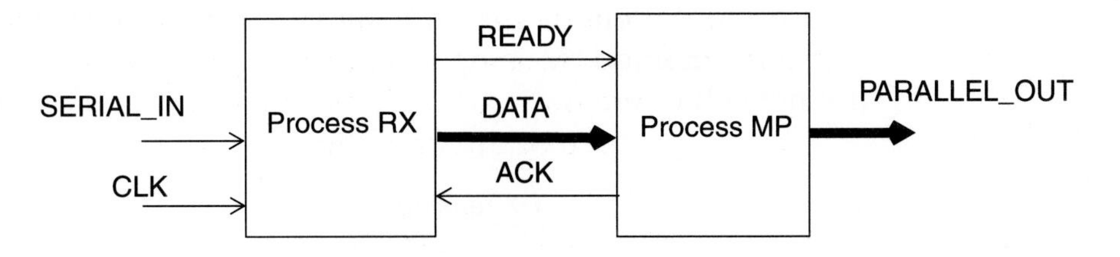

Figure 4-7 Two interacting processes.

The behavior for these two interacting processes is expressed in the following design description.

```
entity INTERACTING is
  port (SERIAL_IN, CLK: in BIT;
        PARALLEL_OUT: out BIT_VECTOR (0 to 7));
end INTERACTING;

architecture PROCESSES of INTERACTING is
  signal READY, ACK: BIT;
  signal DATA: BIT_VECTOR (0 to 7);
begin
  RX: process
  begin
    READ_WORD (SERIAL_IN, CLK, DATA);
      -- READ_WORD is a procedure described elsewhere that reads
      -- the serial data on every clock cycle and converts to a parallel
      -- data in signal DATA. It takes 10 ns to do this.
    READY <= '1';
    wait until ACK = '1';
    READY <= '0';
    wait for 40 ns;
  end process RX;
  MP: process
  begin
    wait for 25 ns;
    PARALLEL_OUT <= DATA;
    ACK <= '1', '0' after 25 ns;
    wait until READY = '1';
  end process MP;
end PROCESSES;
```

The interaction of these two processes via signals READY and ACK is shown in the waveforms in Figure 4-8.

When multiple processes exist within an architecture body, it is possible for more than one process to drive the same signal. In such a case, that signal has multiple drivers (remember that a single process has only one driver for a signal), and the effective value of the signal is obtained by using a resolution function. This is discussed in the next chapter.

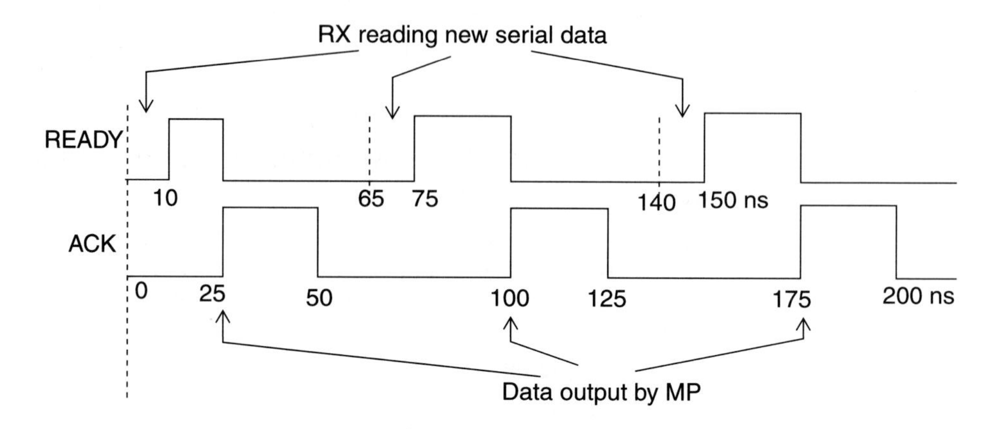

Figure 4-8 Handshaking protocol between the two processes.

4.18 Postponed Processes

In certain situations, it may be useful to trigger a process only at the end of all deltas of a simulation time, instead of executing it in every delta in which the process resumes execution. This can be achieved by marking the process as a postponed process by using the keyword **postponed**, such as:

```
postponed process
   . . .
begin
   . . .
end postponed process;
```

Such a process executes only at the end of all deltas of the simulation time in which an event triggered the process to resume execution.

Here is a simple example.

```
postponed process (A, B)
begin
  Z <= A nor B after 8 ns;
end postponed process;
```

If events occur on signals A or B, say, at times 5 ns, 5 + Δ, 5 + 4Δ, 10 ns, 10 + 3Δ, 16 ns, and 16 + Δ, then the postponed process will execute only after all the deltas at time 5 ns, all the deltas at time 10 ns, and all the deltas at time 16 ns. If the process were not a postponed process (by not using the keyword **postponed**), the process would execute at times 5 ns, 5 + Δ, 5 + 4Δ, 10 ns, 10 + 3Δ, 16 ns, and 16 + Δ.

It is illegal for a postponed process to cause a new delta cycle to occur; that is, a postponed process must not contain a signal assignment or a wait statement that has delta delay. For example, the following process statement is illegal.

```
postponed process (A, B)
begin
  Z <= A nor B;                    -- Delta delay: not allowed.
end postponed process;
```

If signal A had an event at time 5 + 2Δ, this postponed process would execute at the end of all deltas at 5 ns. However, the execution of this statement would cause yet another delta delayed event to occur, which is not allowed.

❑

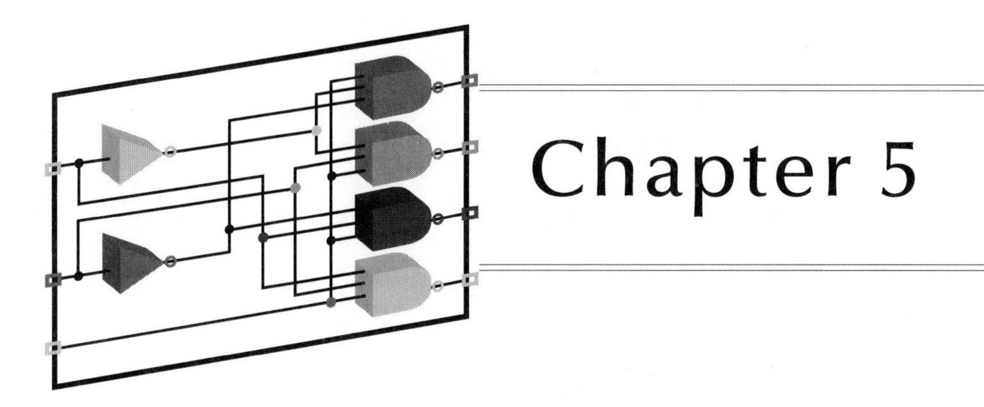

Chapter 5

Dataflow Modeling

This chapter presents techniques for modeling the dataflow of an entity. A dataflow model specifies the functionality of the entity without explicitly specifying its structure. This functionality shows the flow of information through the entity, which is expressed primarily using concurrent signal assignment statements and block statements. This is in contrast to the behavioral style of modeling described in the previous chapter, in which the functionality of the entity is expressed using statements that are executed sequentially. This chapter also describes resolution functions and their usage.

5.1 Concurrent Signal Assignment Statement

One of the primary mechanisms for modeling the dataflow behavior of an entity is using the concurrent signal assignment statement. An example of a dataflow model for a two-input `or` gate, shown in Figure 5-1, follows.

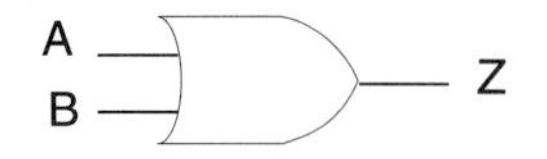

Figure 5-1 An `or` gate.

```
entity OR2 is
   port (signal A, B: in BIT; signal Z: out BIT);
end OR2;

architecture OR2 of OR2 is
begin
   Z <= A or B after 9 ns;
end OR2;
```

The architecture body contains a single concurrent signal assignment statement that represents the dataflow of the `or` gate. The interpretation of this statement is that whenever there is an event (a change of value) on either signal A or B (A and B are signals in the expression for Z), the expression on the right is evaluated, and its value is scheduled to appear on signal Z after a delay of 9 ns. The signals in the expression, A and B, form the "sensitivity list" for the signal assignment statement.

There are two other points to mention about this example. First, the input and output ports have their object class "signal" explicitly specified in the entity declaration. If it were not so, the ports would still have been signals, since this is the default and the only object class that is allowed for ports. The second point to note is that the architecture name and the entity name are the same. This is not a problem, since architecture bodies are secondary units while entity declarations are primary units, and the language allows secondary units to have the same names as primary units.

An architecture body can contain any number of concurrent signal assignment statements. Since they are concurrent statements, the ordering of the statements is not important. Concurrent signal assignment statements are executed whenever events occur on signals that are used in their expressions. An example of a dataflow model for a 1-bit full-adder, whose external view is shown in Figure 5-2, is presented next.

```
entity FULL_ADDER is
   port (A, B, CIN: in BIT; SUM, COUT: out BIT);
end FULL_ADDER;
```

```
architecture FULL_ADDER of FULL_ADDER is
begin
  SUM <= A xor B xor CIN after 15 ns;
  COUT <= (A and B) or (B and CIN) or (CIN and A) after 10 ns;
end FULL_ADDER;
```

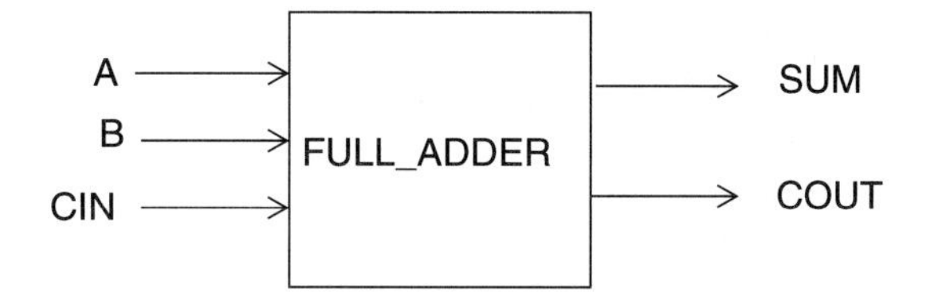

Figure 5-2 External view of a 1-bit full-adder.

Two signal assignment statements are used to represent the dataflow of the FULL_ADDER entity. Whenever an event occurs on signals A, B, or CIN, expressions of both statements are evaluated, and the value to SUM is scheduled to appear after 15 ns, while the value to COUT is scheduled to appear after 10 ns. The `after` clause models the delay of the logic represented by the expression.

Contrast this with the statements that appear inside a process statement. Statements within a process are executed sequentially, while statements in an architecture body are all concurrent statements and order-independent. A process statement is itself a concurrent statement. What this means is that if there were any concurrent signal assignment statements and process statements within an architecture body, the order of these statements also would not matter.

5.2 Concurrent versus Sequential Signal Assignment

In the previous chapter, we saw that signal assignment statements can also appear within the body of a process statement. Such statements are called *sequential* signal assignment statements, while signal assignment statements that appear outside of a process are called *concurrent* signal assignment statements. Concurrent signal assignment statements are event-triggered, that is, they are executed whenever there is an event on a signal that appears in its expression, while sequential signal assignment statements are not event-triggered and are executed in sequence in relation to the other sequential statements that appear within the process. To further understand the difference between these two kinds of signal assignment statements, consider the following two architecture bodies.

```
architecture SEQ_SIG_ASG of FRAGMENT1 is
  -- A, B and Z are signals.
begin
  process (B)
  begin      -- Following are sequential signal assignment statements:
    A <= B;
    Z <= A;
  end process;
end;

architecture CON_SIG_ASG of FRAGMENT2 is
begin      -- Following are concurrent signal assignment statements:
  A <= B;
  Z <= A;
end;
```

In architecture SEQ_SIG_ASG, the two signal assignments are sequential signal assignments. Therefore, whenever signal B has an event (say, at time *T*), the first signal assignment statement is executed and then the second, both in zero time. However, signal A is scheduled to get its new value of B only at time $T+\Delta$ (the delta delay is implicit), and Z is scheduled to be assigned the old value of A (not the value of B) at time $T+\Delta$ also.

In architecture CON_SIG_ASG, the two statements are concurrent signal assignment statements. When an event occurs on signal B (say, at time *T*), signal A gets the value of B after delta delay, that is, at time $T+\Delta$. When simulation time advances to $T+\Delta$, signal A will get its new value, and this event on A (assuming there is a change of value on signal A) will trigger the second signal assignment statement, which will cause the new value of A to be assigned to Z after another delta delay, that is, at time $T+2\Delta$. The delta delay model is explored in more detail in the next section.

Aside from the previous difference, the concurrent signal assignment statement is identical to the sequential signal assignment statement in terms of its behavior.

For every concurrent signal assignment statement, there is an equivalent process statement with the same semantic meaning. The concurrent signal assignment statement:

```
CLEAR <= RESET or PRESET after 15 ns;
  -- RESET and PRESET are signals.
```

is equivalent to the following process statement.

```
process
begin
  CLEAR <= RESET or PRESET after 15 ns;
  wait on RESET, PRESET;
end process;
```

An identical signal assignment statement (now a sequential signal assignment) appears in the body of the process statement along with a `wait` statement whose sensitivity list comprises signals used in the expression of the concurrent signal assignment statement.

A concurrent signal assignment statement can be marked as postponed, such as:

```
CLEAR <= postponed RESET or PRESET after 15 ns;
```

The semantics of such a statement is equivalent to the semantics of its equivalent process statement, which is a postponed process.

5.3 Delta Delay Revisited

In a signal assignment statement, if no delay is specified or a delay of 0 ns is specified, a delta delay is assumed. Delta delay is an infinitesimally small amount of time. It is not a real-time delta and does not cause real simulation time to change. The delta delay mechanism provides for ordering of events on signals that occur at the same simulation time. Consider the circuit shown in Figure 5-3 and the corresponding model.

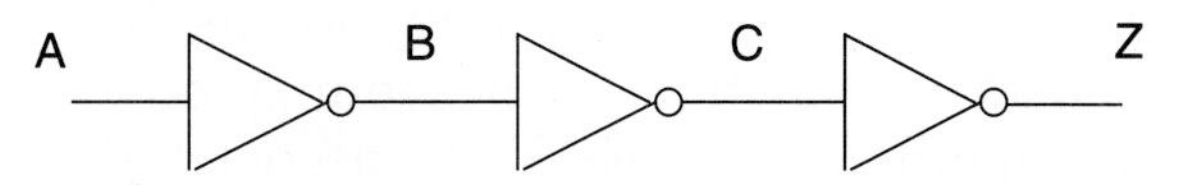

Figure 5-3 Three inverting buffers in series.

```
entity FAST_INVERTER is
   port (A: in BIT; Z: out BIT);
end;

architecture DELTA_DELAY of FAST_INVERTER is
   signal B, C: BIT;
begin    -- Following statements are order-independent:
   Z <= not C;                           -- signal assignment #1
   C <= not B;                           -- signal assignment #2
   B <= not A;                           -- signal assignment #3
end;
```

The three signal assignments in the FAST_INVERTER entity use delta delays. When an event occurs on signal A, say at 20 ns, the third signal assignment is triggered, which causes signal B to get the inverted value of A at 20 ns+1Δ. When time advances to 20 ns+1Δ, signal B changes. This triggers the second signal assignment, causing signal C to get the inverted value of B after another delta delay, that is, at 20 ns+2Δ. When simulation time advances to 20 ns+2Δ, the first signal assignment is triggered, causing Z to get a new value at time 20 ns+3Δ. Even though the real simulation time stayed at 20 ns, Z was updated with the correct value through a sequence of delta-delayed events. This sequence of waveforms is shown in Figure 5-4.

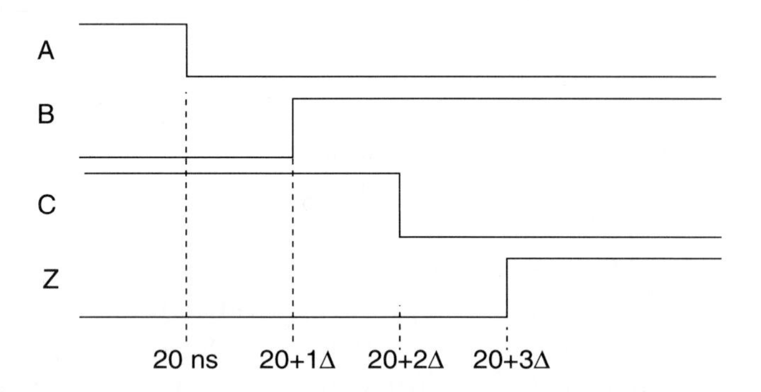

Figure 5-4 Delta delays in concurrent signal assignment statements.

A typical VHDL simulator maintains a list of events that is to occur in an event queue. The events in this queue are ordered not only on the real simulation time, but also on each delta delay. Figure 5-5 shows a snapshot of an event queue in a VHDL simulator during simulation. Each event has associated with it a list of signal-value pairs that are to be scheduled. For example, the value '0' is to be assigned to signal Z when simulation time advances to 10 ns+2Δ. Any postponed process is evaluated only at the end of all the deltas of a simulation time, that is, in the marked shaded regions. For example, if a postponed process activates at time 10+1Δ, the process will actually resume execution only after events at time 10+2Δ is processed.

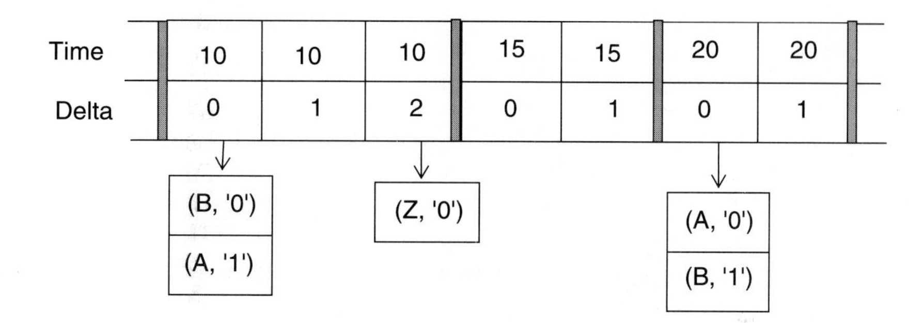

Figure 5-5 An event queue in a VHDL simulator.

Let us consider another example of delta delay. The dataflow model for the RS latch shown in Figure 5-6 appears next.

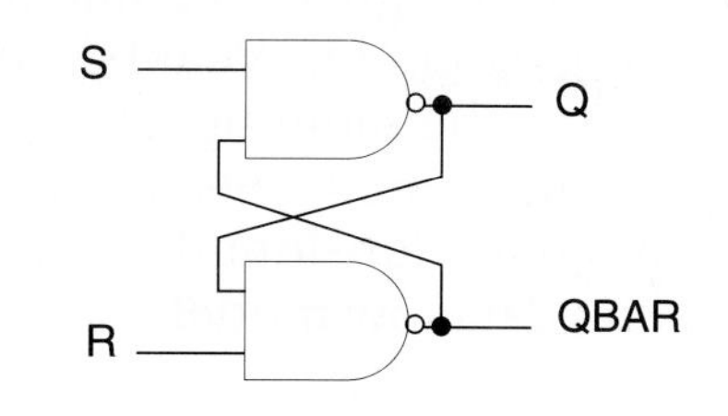

Figure 5-6 An RS latch.

```
entity RS_LATCH is
  port (R, S: in BIT; Q: buffer BIT;
     QBAR: buffer BIT);
end RS_LATCH;

architecture DELTA of RS_LATCH is
begin
  QBAR <= R nand Q;
  Q <= S nand QBAR;
end DELTA;
```

Assume both R and S have value '1' and Q and QBAR are at '1' and '0', respectively. Say signal R changes from '1' to '0' at 5 ns. Figure 5-7 shows the sequence of events that occur as a result. After two delta delays, the circuit stabilizes with the final values of Q and QBAR being '0' and '1', respectively.

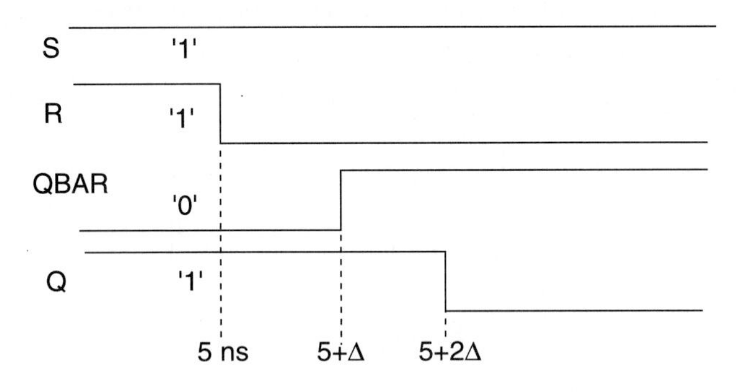

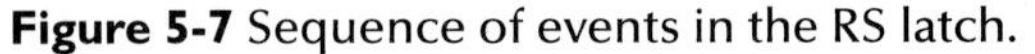

Figure 5-7 Sequence of events in the RS latch.

5.4 Multiple Drivers

Each concurrent signal assignment statement creates a driver for the signal being assigned. What happens when there is more than one assignment to the same signal? In this case, the signal has more than one driver, and a mechanism is needed to compute the effective value of the signal. Consider the circuit shown in Figure 5-8 and its corresponding dataflow model.

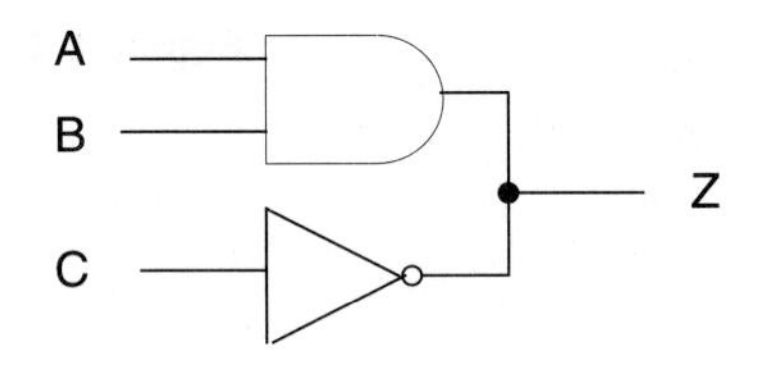

Figure 5-8 Two drivers driving signal Z.

```
entity TWO_DR_EXAMPLE is
   port (A, B, C: in BIT; Z: out BIT);
end TWO_DR_EXAMPLE;

architecture NOT_LEGAL of TWO_DR_EXAMPLE is
begin
   Z <= A and B after 10 ns;
   Z <= not C after 5 ns;
```

```
end ;                     -- Effective value for signal Z has to be
                          -- determined: this is not a legal VHDL model.
```

In this example, there are two gates driving the output signal Z. How is the value of Z determined? It is determined by using a user-defined *resolution function* that considers the values of both the drivers for Z and determines the effective value. Consider the following architecture body.

```
architecture NO_ENTITY of DUMMY is
begin
  Z <= '1' after 2 ns, '0' after 5 ns, '1' after 10 ns;
  Z <= '0' after 4 ns, '1' after 5 ns, '0' after 20 ns;
  Z <= '1' after 10 ns, '0' after 20 ns;
end NO_ENTITY;
```

In this case, there are three drivers for signal Z. Each driver has a sequence of transactions where each transaction defines the value to appear on the signal and the time at which it is to appear. The resolution function resolves the value for the signal Z from the current value of each of its drivers. This is shown pictorially in Figure 5-9.

The value of each driver is an input to the resolution function and, based on the computation performed within the resolution function, the value returned by this function becomes the resolved value for the signal. The resolution function is user-written and may perform any function. The function is not restricted to perform a wired-and or wired-or operation; for example, it could be used to count the number of events on a signal.

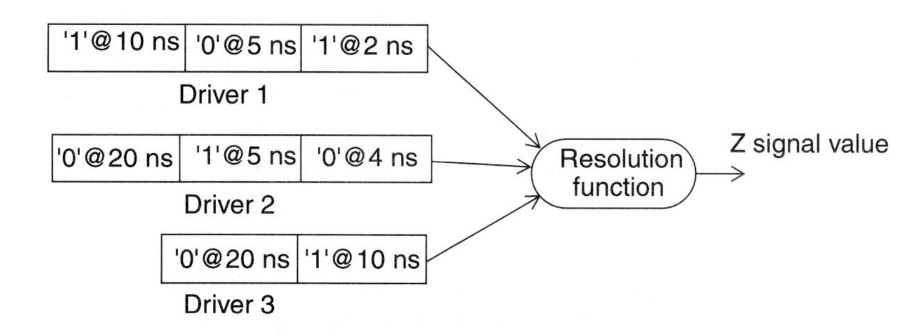

Figure 5-9 Resolving signal drivers.

A signal with more than one driver must have a resolution function associated with it; otherwise, it is an error. Such a signal is called a *resolved* signal. A resolution function is associated with a signal by specifying its name in the signal declaration. For example,

```
signal BUSY: WIRED_OR BIT;
```

is one way of associating the resolution function WIRED_OR with the signal BUSY. No arguments can be specified, since the arguments for the function are the current values of all the drivers for that signal. The TWO_DR_EXAMPLE entity previously described is therefore incorrect; a resolution function must be specified for port Z, such as:

```
port (A,B, C: in BIT; Z: out WIRED_OR BIT);
```

Another way is by declaring a *resolved subtype*, that is, by including the name of the resolution function in the subtype declaration and then declaring the signal to be of that subtype. For example,

```
subtype RESOLVED_BIT is WIRED_OR BIT;
signal BUSY: RESOLVED_BIT;
```

The resolved signal Z in the TWO_DR_EXAMPLE entity can now be specified as:

```
port (A, B, C: in BIT; Z: out RESOLVED_BIT);
```

The semantics of when a resolution function is invoked are as follows. Whenever an event occurs on any source of a resolved signal, the resolution function associated with that signal is called with the values of all its drivers. The return value from the resolution function becomes the value for the resolved signal. In the example of architecture NO_ENTITY, the resolution function is invoked at time 2 ns with driver values '1', '0', and '0' (drivers 2 and 3 have '0' because that is assumed to be the initial value of Z). The function WIRED_OR is performed, and the resulting resolved value of '1' is assigned to Z at 2 ns. Signal Z is scheduled to have another event at 4 ns, at which time the driver values '1', '0', and '0' are passed to the resolution function, which returns the value of '1' for signal Z. At time 5 ns, the driver values, '0', '1', and '0' are passed to the resolution function which returns the value '1'. At 10 ns, the driver values '1', '1', and '1' are passed to the resolution function. Finally, at time 20 ns, the driver values '1', '0', and '0' are passed to the resolution function to determine the effective value for signal Z, which is '1'.

The resolution function has only one input parameter, a constant one-dimensional unconstrained array. The array element type and the return type of the function are the same type as the signal. The function typically computes a value from the various driver values, each element of the input array corresponding to one of the driver values. It should be noted that the identity of the driver is lost in the input array, that is, there is no way of knowing which driver is associated with which element of the input array. Here is an example of a WIRED_OR function that can be used as a resolution function.

```
function WIRED_OR (INPUTS: BIT_VECTOR) return BIT is
begin
  for J in INPUTS'RANGE loop
    if INPUTS(J) = '1' then
      return '1';
    end if;
  end loop;
  return '0';
end WIRED_OR;
```

There is nothing special about the syntax for a resolution function. It is identical to that of any other function (functions are described in Chapter 8). A function is recognized as a resolution function if it is associated with a signal in the signal declaration or a subtype declaration. In the previous example, we introduced a predefined attribute 'RANGE of an array object. This attribute returns the range of the elements of the specified array object. For example, if there are four drivers when the WIRED_OR function is called, INPUTS'RANGE returns the range "0 **to** 3".

If a signal KCQ is of type STD_ULOGIC (type STD_ULOGIC is declared in package STD_LOGIC_1164) and has multiple drivers, then a resolution function must be used. A resolution function called RESOLVED is predefined in the STD_LOGIC_1164 package. Here is a possible declaration for signal KCQ.

```
signal KCQ: RESOLVED STD_ULOGIC;
```

Alternately, the resolved subtype STD_LOGIC (subtype STD_LOGIC is declared in package STD_LOGIC_1164) can be used leading to a more compact signal declaration for KCQ.

```
signal KCQ: STD_LOGIC;
```

Any multiple drivers for KCQ will imply a call to the function RESOLVED to determine the effective value for signal KCQ.

Drivers are also created for signals that are assigned within a process. The one difference is that irrespective of how many times a signal is assigned a value inside a process, there is only one driver for that signal in that process. Therefore, each process will create at most one driver for a signal. If a signal is assigned a value using multiple concurrent signal assignment statements (which can only appear outside a process), an equal number of drivers are created for that signal.

The following two sections present two other forms of the concurrent signal assignment statement: the conditional signal assignment statement and the selected signal assignment statement.

5.5 Conditional Signal Assignment Statement

The conditional signal assignment statement selects different values for the target signal based on the specified, possibly different, conditions. (It is like an `if` statement.) A typical syntax for this statement is:

```
target-signal <=   [ waveform-elements when condition else ]
                   [ waveform-elements when condition else ]
                   . . .
                   waveform-elements [ when condition ] ;
```

The semantics of this concurrent statement are as follows. Whenever an event occurs on a signal used in either any of the waveform expressions (recall that a waveform expression is the value expression in a waveform element) or any of the conditions, the conditional signal assignment statement is executed by evaluating the conditions one at a time. For the first true condition found, the corresponding value (or values) of the waveform is scheduled to be assigned to the target signal. For example,

```
Z <=   IN0 after 10 ns when S0 = '0' and S1 = '0' else
       IN1 after 10 ns when S0 = '1' and S1 = '0' else
       IN2 after 10 ns when S0 = '0' and S1 = '1' else
       IN3 after 10 ns;
```

In this example, the statement is executed any time an event occurs on signals IN0, IN1, IN2, IN3, S0, or S1. The first condition (S0 = '0' **and** S1 = '0') is checked; if false, the second condition (S0 = '1' **and** S1 = '0') is checked; if false, the third condition is checked; and so on. Assuming S0 = '0' and S1 = '1', then the value of IN2 is scheduled to be assigned to signal Z after 10 ns.

For a given conditional signal assignment statement, there is an equivalent process statement that has the same meaning. Such a process statement has exactly one `if` statement and one `wait` statement within it. The signals in the sensitivity list for the `wait` statement are the signals in all the waveform expressions and the signals referenced in all the conditions. The equivalent process statement for these conditional signal assignment statement example is:

```
process
begin
  if S0 = '0' and S1 = '0' then
    Z <= IN0 after 10 ns;
  elsif S0 = '1' and S1='0' then
    Z <= IN1 after 10 ns;
  elsif S0 = '0' and S1 = '1' then
    Z <= IN2 after 10 ns;
  else
    Z <= IN3 after 10 ns;
  end if;
  wait on IN0, IN1, IN2, IN3, S0, S1;
end process;
```

5.6 Selected Signal Assignment Statement

The selected signal assignment statement selects different values for a target signal based on the value of a select expression (it is like a `case` statement). A typical syntax for this statement is:

```
with expression select                          -- This is the select expression.
  target-signal <=     waveform-elements when choices ,
                       waveform-elements when choices ,
                       . . .
                       waveform-elements when choices ;
```

The semantics of a selected signal assignment statement is very similar to that of the conditional signal assignment statement. Whenever an event occurs on a signal in the select expression or on any signal used in any of the waveform expressions, the statement is executed. Based on the value of the select expression that matches the choice value specified, the value (or values) of the corresponding waveform is scheduled to be assigned to the target signal. Note that the choices are not evaluated in sequence. All possible values of the select expression must be covered by the choices exactly once. Values not covered explicitly may be covered by an "others" choice, which covers all such values. The choices may also be an "or" of several values or may be specified as a range of values.

Here is an example of a selected signal assignment statement.

```
type OP is (ADD, SUB, MUL, DIV);
signal OP_CODE: OP;
. . .
with OP_CODE select
   Z <=   A + B after ADD_PROP_DLY when ADD,
          A – B after SUB_PROP_DLY when SUB,
          A * B after MUL_PROP_DLY when MUL,
          A / B after DIV_PROP_DLY when DIV;
```

In this example, whenever an event occurs on signals OP_CODE, A, or B, the statement is executed. Assuming the value of the select expression OP_CODE is SUB, the expression "A – B" is computed, and its value is scheduled to be assigned to signal Z after SUB_PROP_DLY time.

For every selected signal assignment statement, there is an equivalent process statement with the same semantics. In the equivalent process statement, there is one `case` statement that uses the select expression to branch. The list of signals in the sensitivity list of the `wait` statement comprises all signals in the select expression and in the waveform expressions. The equivalent process statement for the previous example is:

```
process
begin
   case OP_CODE is
      when ADD      => Z <= A + B after ADD_PROP_DLY;
      when SUB      => Z <= A – B after SUB_PROP_DLY;
      when MUL      => Z <= A * B after MUL_PROP_DLY;
      when DIV      => Z <= A / B after DIV_PROP_DLY;
   end case;
   wait on OP_CODE, A, B;
end process;
```

5.7 The UNAFFECTED Value

It is possible to assign a value of **unaffected** to a signal in a concurrent signal assignment statement. Such an assignment causes no change to the driver for the target signal. Here are some examples.

```
type STATE_TYPE is (RESET, APPLY, WAITS, HOLD, RECEIVE);
signal NEXT_STATE: STATE_TYPE;
. . .
with NEXT_STATE select
  ZRX <=      B"0001" when APPLY,
              B"0010" when WAITS,
              B"0100" when RESET,
              unaffected when others;

MARK_FLAG <=        BKDET after 5 ns when STROBE= '0' else
                    unaffected;
```

In the first concurrent signal assignment, when NEXT_STATE has the value HOLD or RECEIVE, the value **unaffected** is assigned to signal ZRX. This does not cause any change to the driver, that is, signal ZRX retains its old value. In the second signal assignment, when STROBE has a value other than a '0', MARK_FLAG retains its old value, since an assignment of value **unaffected** does not cause any change to the driver of MARK_FLAG.

It is easier to understand the effects of the **unaffected** value by looking at its equivalent representation in a process statement. An assignment of **unaffected** to a signal translates to a null statement in the equivalent process for the concurrent signal assignment statement. The equivalent process statement for the second example described earlier is:

```
process
begin
  if STROBE = '0' then
    MARK_DET <= BKDET after 5 ns;
  else
    null;                                    -- From the value unaffected.
  end if;
  wait on STROBE, BKDET;
end process;
```

5.8 Block Statement

A block statement is a concurrent statement. It can be used for three major purposes:

1. To disable signal drivers by using guards
2. To limit scope of declarations, including signal declarations
3. To represent a portion of a design

A block statement itself has no execution semantics but provides additional semantics for statements that appear within it. The syntax of a block statement is:

```
block-label : block [ ( guard-expression ) ] [ is ]
      [ block-header ]
      [ block-declarations ]
begin
   concurrent-statements
end block [ block-label ] ;
```

The *block-header*, if present, describes the interface of the block statement to its environment and is discussed in greater detail in Chapter 10. Any declarations appearing within the block are visible only within the block, that is, between **block . . . end block**. Any number of concurrent statements can appear within a block, possibly none. Block statements can be nested since a block statement is itself a concurrent statement. The block label present at the beginning of the block statement is necessary; however, the label appearing at the end of the block statement is optional and, if present, must be the same as the one used at the beginning of the block.

If a *guard-expression* appears in a block statement, there is an implicit signal called *GUARD* of type BOOLEAN declared within the block. The value of the GUARD signal is always updated to reflect the value of the guard expression. The guard expression must be of type BOOLEAN. Signal assignment statements appearing within the block statement can use this GUARD signal to enable or disable their drivers. Here is an example of a gated inverter.

```
B1: block (STROBE = '1')
begin
   Z <= guarded not A;
end block B1;
```

The signal GUARD implicitly declared within block B1 has the value of the expression (STROBE = '1'). The keyword **guarded** can optionally be used with concurrent signal assignment statements within a block statement. This keyword implies that when the value of the GUARD signal is true (that is, the guard expression evaluates to true), the value of the expression "**not** A" is assigned to the target signal Z. If the GUARD is false, events on A do not affect the value of signal Z. That is, the driver to Z for this signal assignment statement is disabled, and signal Z retains its previous value. The block statement is very useful in modeling hardware elements that trigger on certain events, for example, flip-flops and clocked logic.

The only concurrent statements whose semantics are affected by the enclosing block statement are the *guarded assignments*, that is, the signal assignment statements that use the **guarded** option. The modified meaning is as follows. Whenever an event occurs (an event is a change of value) on any signal used in the expression of a guarded assignment and the signal GUARD has the value TRUE, or if signal GUARD changes from FALSE to TRUE, the signal assignment statement is executed, and the target signal is scheduled to get a new value. If the value of the guard expression is false, the value of the target signal is unchanged.

Every guarded assignment has an equivalent process statement. Here is an example.

```
BG: block (guard-expression)
  signal SIG: BIT;
begin
  SIG <= guarded waveform-elements;
end block BG;

-- The equivalent process statement for the guarded assignment is:
BG: block (guard-expression)
  -- signal GUARD: BOOLEAN;                    -- Implicit.
  signal SIG: BIT;
begin
  -- GUARD <= guard-expression;                -- Implicit.
  process
  begin
    if GUARD then
      SIG <= waveform-elements;
    end if;
    wait on signals-in-waveform-elements, GUARD;
  end process;
end block BG;
```

The signal GUARD, even though implicitly declared, can be used explicitly within a block statement. For example,

```
B2: block (CLEAR = '0' and PRESET = '1')
begin
  Q <= '1' when not GUARD else '0';
end block B2;
```

In this example, the signal assignment in the block statement is not a guarded assignment; therefore, the driver to signal Q is never disabled. However, the value of Q is determined by the value of the GUARD signal because of its explicit use in the signal assignment statement. The value of the GUARD signal corresponds to the value of the guard expression " CLEAR = '0' **and** PRESET = '1' ". This signal assignment statement is therefore executed any time an event occurs on signal GUARD.

It is also possible to explicitly declare a signal called GUARD, define an expression for it, and then use it within a guarded assignment. Here is an example.

```
B3: block
  signal GUARD:  BOOLEAN;
begin
  GUARD <= CLEAR = '0' and PRESET = '1';
  Q <= guarded DIN;
end block B3;
```

Here is another example that models a rising-edge triggered D flip-flop.

```
entity D_FLIP_FLOP is
  port (D, CLK: in BIT; Q, QBAR: out BIT);
end D_FLIP_FLOP;

architecture DFF of D_FLIP_FLOP is
begin
  L1: block (CLK = '1' and not CLK'STABLE)
    signal TEMP: BIT;
  begin
    TEMP <= guarded D;
    Q <= TEMP;
    QBAR <= not TEMP;
  end block L1;
end DFF;
```

The guard expression uses a predefined attribute called 'STABLE. CLK'STABLE is a new signal of type BOOLEAN, which is true as long as signal CLK has not had any event in the current delta time. This guard expression implies a rising clock edge. The scope of the signal TEMP declared in block L1 is restricted to be within the block. Of the three signal assignments, only the first one is a guarded assignment and, hence, controlled by the guard expression. The other two signal assignments are not controlled by the guard expression and are triggered purely on events occurring on signals in their expressions. When a rising clock edge appears on the signal CLK, say at time T, the value of D is assigned to signal TEMP after delta delay, that is, at time $T+\Delta$. If the value of TEMP is different from its previous value, the assignments to Q and QBAR will be triggered, causing these signals to get new values after another delta delay, that is, at time $T+2\Delta$.

Other uses of the block statement are described in Chapter 10.

5.9 Concurrent Assertion Statement

Sequential assertion statements were discussed in the previous chapter. A concurrent assertion statement has exactly the same syntax as a sequential assertion statement. An assertion statement is a concurrent statement by virtue of its place of appearance within the model. If it appears inside a process, it is a sequential assertion statement and is executed sequentially with respect to the other statements in the process; if it appears outside a process, it is a concurrent assertion statement. The semantics of a concurrent assertion statement are as follows. Whenever an event occurs on a signal in the Boolean expression of the assertion statement, the statement is executed.

Here is an example of a concurrent assertion statement used in an SR flip-flop model. It makes a check to ensure that the input signals R and S are never simultaneously zero.

```
library IEEE;
use IEEE.STD_LOGIC_1164.all;
entity SR is
   port (S, R: in STD_LOGIC; Q, NOTQ: out STD_LOGIC);
end SR;

architecture SR_ASSERT of SR is
begin
```

```
    assert not (S = '0' and R = '0')
      report "Not valid inputs: R and S are both low"
      severity ERROR;
    -- Rest of model for SR flip-flop here.
  end SR_ASSERT;
```

Any time an event occurs on either signal S or R, the assertion statement is executed and the Boolean expression checked. If false, the report message is printed, and the severity level is reported to the simulator for appropriate action.

For every concurrent assertion statement there is an equivalent process statement with the same semantic meaning. The equivalent process statement for the previous example of the concurrent assertion statement is:

```
process
begin
  assert not (S = '0' and R = '0')
    report "Not valid inputs: R and S are both low"
    severity ERROR;
  wait on S, R;
end process;
```

A concurrent assertion statement can be marked as a postponed concurrent assertion statement, such as:

```
postponed assert not (S = '0' and R = '0')
    report "Not valid inputs: R and S are both low"
    severity ERROR;
```

The semantics of such a statement is equivalent to the semantics of its equivalent process statement, which is a postponed process.

5.10 Value of a Signal

A signal gets its value from its drivers. Every concurrent signal assignment statement creates a driver for the target signal. All signal assignments in a process that assign to a particular signal create one driver for that signal.

In a given VHDL description, if a signal has more than one driver, a resolution function is necessary. This function is called with the current value of all the drivers for a signal, and the return value of the function becomes the effective value of the signal.

Here is an example.

```
library IEEE;          -- Library and use clauses can also be associated
                       -- with an architecture body.
use IEEE.STD_LOGIC_1164.all;
architecture DRIVERS of INTERCON is
   signal FRMERR, PARERR: STD_LOGIC := 'Z';
   -- Type STD_LOGIC is defined in package STD_LOGIC_1164 as:
   -- subtype STD_LOGIC is RESOLVED STD_ULOGIC;
   signal XFSB: BIT;
begin
   P1: process ( . . . )
      . . .
   begin
      FRMERR <= . . .
      XFSB <= . . .
      . . .
      XFSB <= . . .
   end process P1;
   P2: process ( . . . )
      . . .
   begin
      PARERR <= . . .
      FRMERR <= . . .
      . . .
   end process P2;
   PARERR <= . . .
end DRIVERS;
```

In this example, signal FRMERR has two drivers, one from process P1 and one from process P2. Therefore, a resolution function is needed, and this (function RESOLVED) is specified in the signal declaration of FRMERR (STD_LOGIC is a resolved subtype with the resolution function RESOLVED). Similarly, signal PARERR has two drivers and a resolution function is specified in its signal declaration. However, signal XFSB has only one driver from process P1, and therefore needs no resolution function.

The effective value for signals FRMERR and PARERR is the return value of the resolution function, while the effective value of signal XFSB is the current value of the driver itself. Note that a resolution function could be specified for signal XFSB as well, such as in the following declaration.

```
signal XFSB: PULL_UP BIT;
```

In this case, the value of its one driver is the input to the resolution function, and the return value of the function is the effective value of signal XFSB. Clearly, if the function PULL_UP were simply passing on the input value to the output in the case of one driver, then specifying a resolution function simply adds to the simulation overhead since an extra function call is involved.

The recommended practice, therefore, is not to specify a resolution function if a signal has only one driver.

The initial value of a signal, that is, its value at start of simulation, is determined in a similar way from the value of its drivers. The initial value of a driver of a signal is the initial value as specified in the signal declaration, either implicitly, for example, using *signal-type*'LEFT for scalars, or explicitly, using ":=" in the signal declaration. If a signal has a resolution function associated with it (it may or may not have more than one driver), the initial value of the signal is computed by passing the initial values of all the drivers to the resolution function, and the return value becomes the initial value of the signal. If a signal does not have a resolution function associated with it (it can only have one driver), the initial value of the signal is the same as the initial value of the driver.

In the architecture body DRIVERS described earlier, the initial value of the two drivers for FRMERR is 'Z' (the explicitly specified initial value). These two values are passed to the resolution function RESOLVED. The return value of the function is the effective initial value of FRMERR. This value may or may not be 'Z', depending on the computation performed by the resolution function. The initial value of signal XFSB is the initial value of the driver (since there is only one driver), which is the implicit initial value of '0' (this is BIT'LEFT).

❑

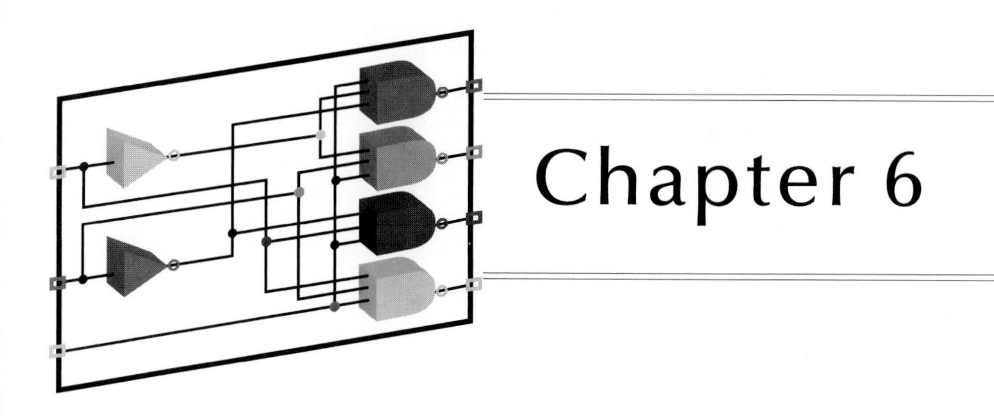

Chapter 6

Structural Modeling

This chapter describes the structural style of modeling. An entity is modeled as a set of components connected by signals, that is, as a netlist. The behavior of the entity is not explicitly apparent from its model. The component instantiation statement is the primary mechanism used for describing such a model of an entity.

6.1 An Example

Consider the circuit shown in Figure 6-1 and its VHDL structural model.

```
entity GATING is
   port (A, CK, MR, DIN: in BIT; RDY, CTRLA: out BIT);
end GATING;

architecture STRUCTURE_VIEW of GATING is
   component AND2
      port (X, Y: in BIT; Z: out BIT);
   end component;
```

```
    component DFF
        port (D, CLOCK: in BIT; Q, QBAR: out BIT);
    end component;
    component NOR2
        port (DA, DB: in BIT; DZ: out BIT);
    end component;
    signal S1, S2: BIT;
begin
    D1: DFF port map (A, CK, S1, S2);
    A1: AND2 port map (S2, DIN, CTRLA);
    N1: NOR2 port map (S1, MR, RDY);
end STRUCTURE_VIEW;
```

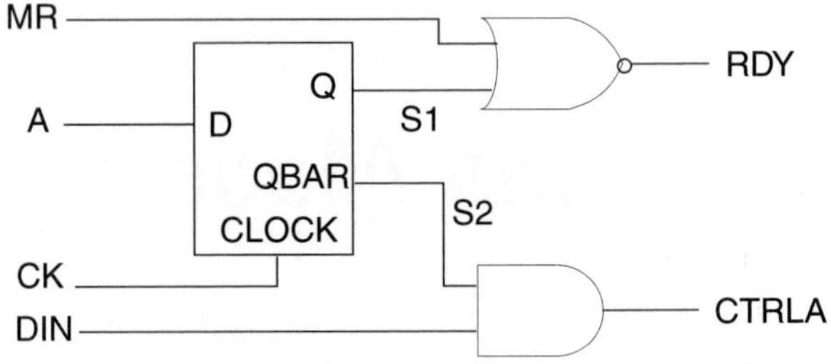

Figure 6-1 A circuit generating control signals.

Three components, AND2, DFF, and NOR2, are declared. These components are instantiated in the architecture body via three component instantiation statements, and the instantiated components are connected to each other via signals S1 and S2. The component instantiation statements are concurrent statements, and their order of appearance in the architecture body is therefore not important. A component can, in general, be instantiated any number of times. However, each instantiation must have a unique component label; as an example, A1 is the component label for the AND2 component instantiation.

6.2 Component Declaration

A component instantiated in a structural description must first be declared using a component declaration. A component declaration declares the name and the interface of a component. The interface specifies the mode and the type of ports. The syntax of a simple form of component declaration is:

```
component component-name [ is ]
   [ port ( list-of-interface-ports ) ; ]
end component [ component-name ] ;
```

The *component-name* may or may not refer to the name of an entity already existing in a library. If it does not, it must be explicitly bound to an entity; otherwise, the model cannot be simulated (the model may still be analyzed). The binding information can be specified using a configuration. Configurations are discussed in the next chapter.

The *list-of-interface-ports* specifies the name, mode, and type for each port of the component in a manner similar to that specified in an entity declaration. The names of the ports may also be different from the names of the ports in the entity to which it may be bound (different port names can be mapped in a configuration). In this chapter, we will assume that an entity of the same name as that of the component already exists and that the name, mode, and type of each port matches the corresponding ones in the component. Some examples of component declarations are:

```
component NAND2
   port (A, B: in STD_LOGIC; Z: out STD_LOGIC);
end component;

component MP
   port (CK, RESET, RDN, WRN: in BIT;
            DATA_BUS: inout INTEGER range 0 to 255;
            ADDR_BUS: in BIT_VECTOR(15 downto 0));
end component MP;

component RX
   port (CK, RESET, ENABLE, DATAIN, RD: in BIT;
            DATA_OUT: out INTEGER range 0 to (2**8 - 1);
            PARITY_ERROR, FRAME_ERROR,
            OVERRUN_ERROR: out BOOLEAN);
end component;
```

Component declarations appear in the declarations part of an architecture body. Alternately, they may also appear in a package declaration. Items declared in this package can then be made visible within any architecture body by using the `library` and `use` clauses. For example, consider the entity GATING described in the previous section. A package such as the one shown next may be created to hold the component declarations.

```
package COMP_LIST is
   component AND2
      port (X,Y: in BIT; Z: out BIT);
   end component;
   component DFF
      port (D, CLOCK: in BIT; Q, QBAR: out BIT);
   end component;
   component NOR2
      port (DA, DB: in BIT; DZ: out BIT);
   end component;
end COMP_LIST;
```

Assuming that this package has been compiled into design library DES_LIB, the architecture body can be rewritten as:

```
library DES_LIB;
use DES_LIB.COMP_LIST.all;
architecture STRUCTURE_VIEW of GATING is
   signal S1, S2: BIT;
   -- No need for specifying component declarations here, since they
   -- are made visible to architecture body
   -- using library and use clauses.
begin
   -- The component instantiations here.
end STRUCTURE_VIEW;
```

The advantage of this approach is that the package can now be shared by other design units, and the component declarations need not be specified inside every design unit.

6.3 Component Instantiation

A component instantiation statement defines a subcomponent of the entity in which it appears. It associates the signals in the entity with the ports of that subcomponent. A format of a component instantiation statement is:

component-label : *component-name* [**port map** (*association-list*)] ;

The *component-label* can be any legal identifier and can be considered as the name of the instance. The *component-name* must be the name of a component declared earlier using a component declaration. The *association-list* associates signals in the entity, called *actuals*, with the ports of a component, called *formals*. An actual may be a signal. An actual for an input port may also be an expression. An actual may also be the keyword **open** to indicate a port that is not connected.

There are two ways to perform the association of formals with actuals:

1. Positional association
2. Named association

In positional association, an *association-list* is of the form:

$actual_1, actual_2, actual_3, \ldots, actual_n$

Each actual in the component instantiation is mapped by position with each port in the component declaration. That is, the first port in the component declaration corresponds to the first actual in the component instantiation, the second with the second, and so on. Consider an instance of a NAND2 component.

```
-- Component declaration:
component NAND2
  port (A, B: in STD_LOGIC; Z: out STD_LOGIC);
end component;

-- Component instantiation:
N1: NAND2 port map (S1, S2, S3);
```

N1 is the component label for the current instantiation of the NAND2 component. Signal S1 (which is an actual) is associated with port A (which is a formal) of the NAND2 component, S2 is associated with port B of the NAND2 component, and S3 is associated with port Z. Signals S1 and S2 thus provide the two input values to the NAND2 component, and signal S3 receives the output value from the component. The ordering of the actuals is therefore important.

If a port in a component instantiation is not connected to any signal, the keyword **open** can be used to signify that the port is not connected. For example,

```
N3: NAND2 port map (S1, open, S3);
```

The second input port of the NAND2 component is not connected to any signal. An input port may be left open only if its declaration specifies an initial value. For the previous component instantiation statement to be legal, port B of the component declaration for NAND2 must have an initial value expression.

```
component NAND2
  port (A, B: in STD_LOGIC := '0'; Z: out STD_LOGIC);
  -- Both A and B have an initial value of '0'; however, only
  -- the initial value of B is necessary in this case.
end component;
```

A port of any other mode may be left unconnected as long as it is not an unconstrained array.

In named association, an *association-list* is of the form:

$formal_1$ => $actual_1$, $formal_2$ => $actual_2$, . . . , $formal_n$ => $actual_n$

For example, consider the component NOR2 in the entity GATING described in the first section. The instantiation using named association may be written as:

```
N1: NOR2 port map (DB => MR, DZ => RDY, DA => S1);
```

In this case, the signal MR (an actual), which is declared in the entity port list, is associated with the second port (port DB, a formal) of the NOR2 gate, signal RDY is associated with the third port (port DZ), and signal S1 is associated with the first port (port DA) of the NOR2 gate. In named association, the ordering of the associations is not important since the mapping between the actuals and formals is explicitly specified. An important point to note is that the scope of the formals is restricted to be within the port map part of the instantiation for that component; for example, the formals DA, DB, and DZ of component NOR2 are relevant only within the port map of instantiation of component NOR2.

For either type of association, there are certain rules imposed by the language. First, the types of the formal and actual being associated must be the same. Second, the modes of the ports must conform to the rule that if the formal is readable, so must the actual be; and if the formal is writable, so must the actual be. Since a locally declared signal is considered to be both readable and writable, such a signal may be associated with a formal of any mode. If an actual is a port of mode **in**, it may not be associated with a formal of mode **out** or **inout**; if the actual is a port of mode **out**, it may not be associated with a formal of mode **in** or **inout**; if the actual is a port of mode **inout**, it may be associated with a formal of mode **in**, **out**, or **inout**.

It is important to note that an actual of mode **out** or **inout** indicates the presence of a source for that signal, and therefore, it must be resolved if that signal is multiply driven. A buffer port can never have more than one source; therefore, the only kind of actual that can be associated with a buffer port is another buffer port or a signal that has at most one source.

Another example of a component instantiation is:

```
M1: MICRO port map (UDIN(3 downto 0), WRN, RDN, STATUS(0),
                    STATUS(1), UDOUT(0 to 7), TXDATA);
```

The first actual of the port map refers to a slice of the vectored UDIN signal, WRN and RDN are 1-bit signals, STATUS(0) and STATUS(1) refer to the 0th and 1st element of the STATUS array, UDOUT(0 **to** 7) refers to a slice of the UDOUT vector, and TXDATA refers to an entire vector signal. This example shows that the signals used to interconnect components can also be any of the following:

- Slices
- Vectors
- Array elements

Not only can an actual be any of the above form, even a formal can be a slice, a vector or an array element. This is shown using an example.

```
component CHILD6
  port ( INP: in BIT_VECTOR(0 to 4);
         OUTP: out BIT_VECTOR(3 downto 0));
end component;
signal QB: BIT_VECTOR(4 downto 0);
signal RESET: BIT;
signal SET: BIT_VECTOR(0 to 3);
. . .
INST6B: CHILD6 port map (
                INP(0 to 2) => QB(4 downto 2),
                INP(3) => RESET,
                INP(4) => QB(0),
                OUTP => SET);
```

The first formal, INP(0 **to** 2), is an example of a slice, the second and third formals are examples of array elements and the last formal, OUTP, is the entire vector.

Constraints are passed to a formal port from an actual if the formal port is unconstrained.

```
signal QBUS: BIT_VECTOR (3 downto 0);
signal PAND, QAND: BIT;
signal PBUS: BIT_VECTOR (0 to 8);
```

```
component REDUCE_AND
   port (D: in BIT_VECTOR; AND_OUT: out BIT);
end component;
. . .
P1 : REDUCE_AND port map (QBUS, QAND);
P2 : REDUCE_AND port map (PBUS, PAND);
```

The size of the formal port D has not been explicitly specified in the component declaration for REDUCE_AND. In this case, the size of the formal comes from the actual during instantiation. From the first instantiation, port D has the constraint "3 **downto** 0"; from the second instantiation, formal port D has the constraint "0 **to** 8". The constraints for the formal port are on a "per-instance basis", that is, based on the actual in an instantiation, the formal port gets that constraint which could be different for different instances.

6.4 Other Examples

A complete structural model for a 9-bit parity generator circuit is shown in Figure 6-2.

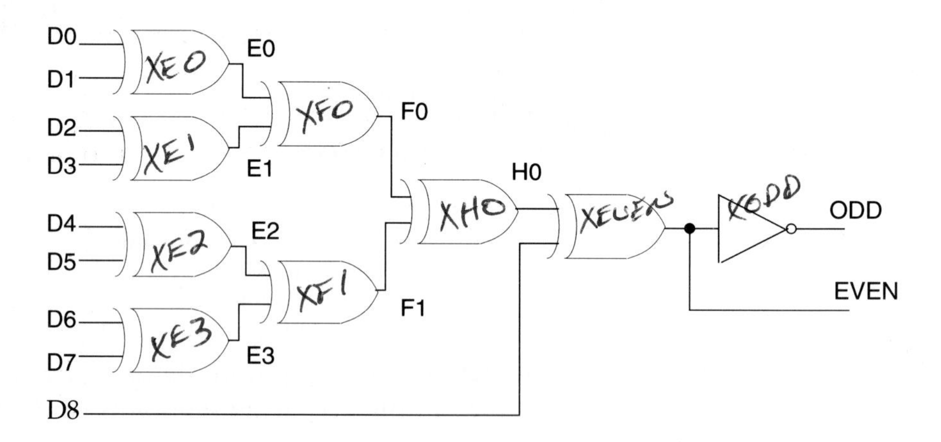

Figure 6-2 A 9-bit parity generator circuit.

```
library IEEE;
use IEEE.STD_LOGIC_1164.all;
entity PARITY_9_BIT is
   port ( D: in STD_LOGIC_VECTOR(8 downto 0);
          ODD: out STD_LOGIC;
```

```
        EVEN: buffer STD_LOGIC);
end PARITY_9_BIT;

architecture PARITY_STR of PARITY_9_BIT is
  component XOR2
    port (A, B: in STD_LOGIC; Z: out STD_LOGIC);
  end component;
  component INV2
    port (A: in STD_LOGIC; Z: out STD_LOGIC);
  end component;
  signal E0, E1, E2, E3, F0, F1, H0: STD_LOGIC;
begin
  XE0: XOR2 port map (D(0), D(1), E0);
  XE1: XOR2 port map (D(2), D(3), E1);
  XE2: XOR2 port map (D(4), D(5), E2);
  XE3: XOR2 port map (D(6), D(7), E3);
  XF0: XOR2 port map (E0, E1, F0);
  XF1: XOR2 port map (E2, E3, F1);
  XH0: XOR2 port map (F0, F1, H0);
  XEVEN: XOR2 port map (H0, D(8), EVEN);
  XODD: INV2 port map (EVEN, ODD);
end PARITY_STR;
```

In this example, port EVEN is of mode **buffer** since the value of this port is being read as well as written to inside the architecture. If this port were declared to be of mode **inout**, external signals defined outside the PARITY_9_BIT design would be able to drive this port, which may not be desired.

An example of a decade counter using J-K flip-flops is shown in Figure 6-3. Its structural model is described next.

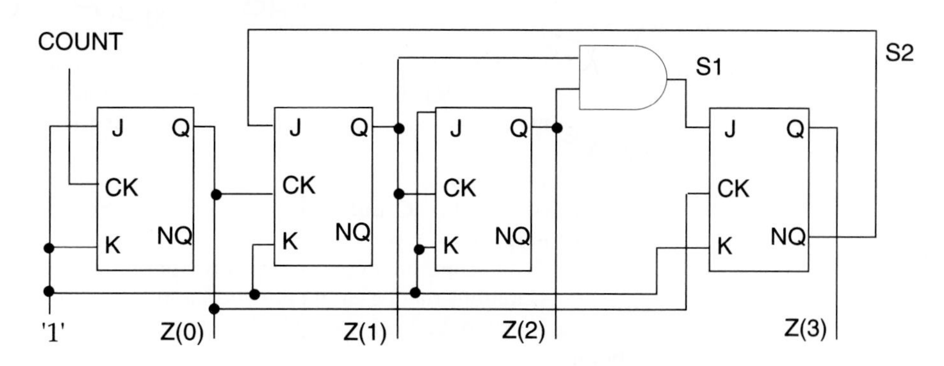

Figure 6-3 A decade counter.

```
entity DECADE_CTR is
  port (COUNT: in BIT; Z: buffer BIT_VECTOR(0 to 3));
end DECADE_CTR;

architecture NET_LIST of DECADE_CTR is
  component JK_FF
    port (J, K, CK: in BIT; Q, NQ: buffer BIT);
  end component;

  component AND_GATE
    port (A, B: in BIT; C: out BIT);
  end component;
  signal S1, S2: BIT;
begin
  A1: AND_GATE port map (Z(2), Z(1), S1);
  JK1: JK_FF port map ('1', '1', COUNT, Z(0), open);
  JK2: JK_FF port map (S2, '1', Z(0), Z(1), open);
  JK3: JK_FF port map ('1', '1', Z(1), Z(2), open);
  JK4: JK_FF port map (S1, '1', Z(0), Z(3), S2);
end NET_LIST;
```

This example illustrates the point that a constant value or, in general, a constant expression can be used as an actual in a port map.

Structural models can be simulated only after the entities that the components represent are modeled and placed in a design library. The lowest-level entities must be behavioral models. The simulation semantics of a component instantiation are best understood by an example. Consider the component instantiation A1 in the previous example. Assume that this instance is bound to an entity with the same name and identical port names. Its equivalent behavioral representation is:

```
A1: block                                 -- A block for each instantiation.
    port (A, B: in BIT; C: out BIT);      -- Ports in component
                                          -- declaration.
    port map (C => S1, A => Z(2), B => Z(1)); -- Association list.
begin
  AND_GATE: block                              -- The entity block.
    port (A, B: in BIT; C: out BIT);           -- Ports of entity.
    port map (A => A, B => B, C => C);         -- Association of
              -- component ports with entity ports.
    -- Declarations that occur in entity declaration and
    -- architecture body of AND_GATE entity appear here.
  begin
    -- Behavior in architecture body for AND_GATE entity.
```

```
      -- For example,
      -- C <= A and B after 10 ns;
    end block AND_GATE;
  end block A1;
```

As shown, a block statement can also have a port list and a port map. The port list specifies the ports through which the block communicates with its external environment. The port map specifies the mapping between the ports and the signals in the block's external environment to which these ports connect. More on this form of block statement is described in Chapter 10.

As a final example, consider the three-bit up-down counter circuit shown in Figure 6-4. It's structural model follows.

```
entity UP_DOWN is
  port (CLK, CNT_UP, CNT_DOWN: in BIT;
        Q0, Q1, Q2: buffer BIT);
end UP_DOWN;

architecture COUNTER of UP_DOWN is
  component JK_FF
    port (J, K, CK: in BIT; Q, QN: buffer BIT);
  end component;
  component AND2
    port (A, B: in BIT; C: out BIT);
  end component;
  component OR2
    port (A, B: in BIT; C: out BIT);
  end component;
  signal S1, S2, S3, S4, S5, S6, S7, S8: BIT;
begin
  JK1: JK_FF port map ('1', '1', CLK, Q0, S1);
  A1: AND2 port map (CNT_UP, Q0, S2);
  A2: AND2 port map (S1, CNT_DOWN, S3);
  O1: OR2 port map (S2, S3, S4);
  JK2: JK_FF port map ('1', '1', S4, Q1, S5);
  A3: AND2 port map (Q1, CNT_UP, S7);
  A4: AND2 port map (S5, CNT_DOWN, S6);
  O2: OR2 port map (S7, S6, S8);
  JK3: JK_FF port map ('1', '1', S8, Q2, open);
end COUNTER;
```

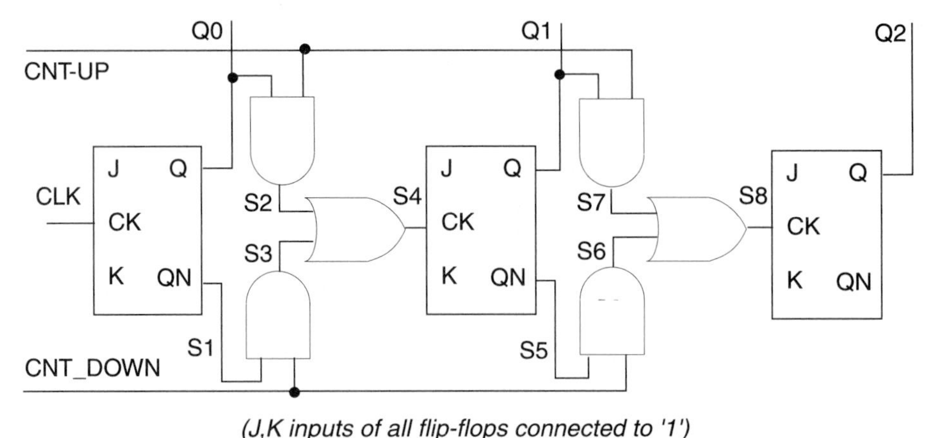

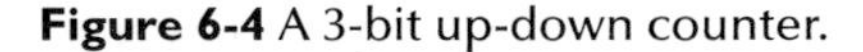

Figure 6-4 A 3-bit up-down counter.

6.5 Resolving Signal Values

If outputs of two components drive a common signal, the value of the signal must be resolved using a resolution function. This is similar to the case of a signal being assigned using more than one concurrent signal assignment statement. For example, consider the circuit shown in Figure 6-5, which shows two `and` gates driving a common signal RS1, which is inverted to produce the result in Z.

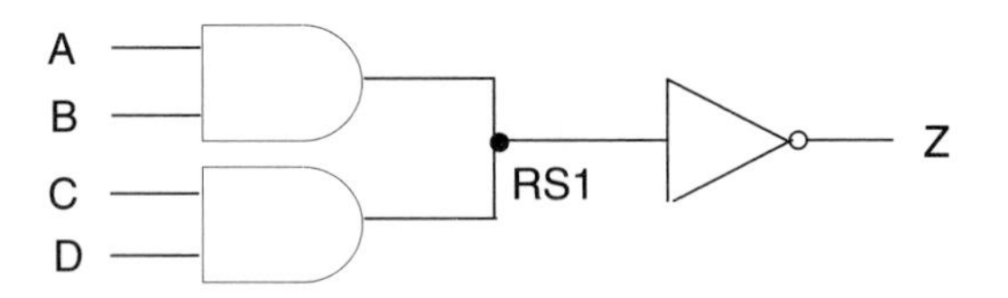

Figure 6-5 Two components driving a common signal.

```
entity DRIVING_SIGNAL is
  port (A, B, C, D: in BIT; Z: out BIT);
end DRIVING_SIGNAL;

-- Assume PULL_UP is the name of a function defined in package
-- RF_PACK that has been compiled into the working library.
use WORK.RF_PACK.PULL_UP;
```

```
architecture RESOLVED of DRIVING_SIGNAL is
  signal RS1: PULL_UP BIT;
  component AND2
    port (IN1, IN2: in BIT; OUT1: out BIT);
  end component;
  component INV
    port (X: in BIT; Y: out BIT);
  end component;
begin
  A1: AND2 port map (A, B, RS1);
  A2: AND2 port map (C, D, RS1);
  I1: INV port map (RS1, Z);
end RESOLVED;
```

The key point here is that even though an assignment to signal RS1 is not being made explicitly using signal assignment statements, the signal RS1 is being driven by two output ports, and therefore must be resolved using a resolution function. In the previous example, the PULL_UP resolution function is associated with signal RS1. This implies that the values of the outputs of the `and` gates are passed through the resolution function before a value is assigned to signal RS1. In general, each **out**, **inout**, and **buffer** port of a component creates a driver for the signal with which it is associated.

❑

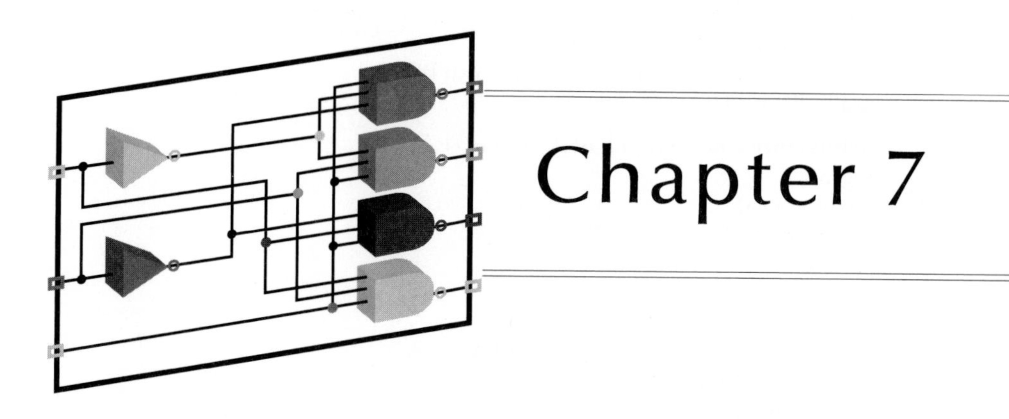

Chapter 7

Generics and Configurations

This chapter introduces generics and provides examples to show how certain types of information can be passed into an entity using generics. Later sections in the chapter discuss the need for configurations and present the two alternate mechanisms provided by the language to configure, namely, the configuration specification and the configuration declaration. Direct instantiation can be used to avoid writing configurations. This feature, along with the delayed binding mechanism, are also described in this chapter.

7.1 Generics

It is often useful to pass certain types of information into a design description from its environment. Examples of such information are rise and fall delays and the size of interface ports. This is accomplished by using generics. Generics of an entity are declared along with its ports in the entity declaration. An example of a generic N-input `and` gate is shown next.

```
entity AND_GATE is
  generic (N: NATURAL);
  port (A: in BIT_VECTOR(1 to N); Z: out BIT);
end AND_GATE;

architecture GENERIC_EX of AND_GATE is
begin
  process (A)
    variable AND_OUT: BIT;
  begin
    AND_OUT := '1';
    for K in 1 to N loop
      AND_OUT := AND_OUT and A(K);
      exit when AND_OUT = '0';
    end loop;
    Z <= AND_OUT;
  end process;
end GENERIC_EX;
```

In this example, the size of the input port has been modeled as a generic. By doing this, we have modeled an entire class of `and` gates with a variable number of inputs using a single behavioral description. The AND_GATE entity may now be used with a different number of input ports in different instantiations.

A generic declares a constant object of mode **in** (that is, the value can only be read) and can be used in the entity declaration and its corresponding architecture bodies. The value of this constant can be specified as a globally static expression in one of the following:

1. Entity declaration
2. Component declaration
3. Component instantiation
4. Configuration specification
5. Configuration declaration

The value of a generic must be determinable at elaboration time, that is, a value for a generic must be explicitly specified at least once using any of the ones mentioned.

The value for a generic may be specified in the entity declaration for an entity, as shown in this example. This is the default value for the generic. It can be overridden by others.

```
entity NAND_GATE is
  generic (M: INTEGER := 2);          -- M models the number of inputs.
```

```
    port (A: in BIT_VECTOR(M downto 1); Z: out BIT);
end NAND_GATE;
```

Two other ways of specifying the value of a generic are in a component declaration and in a component instantiation. The following examples demonstrates these.

```
entity ANOTHER_GEN_EX is
end;

architecture GEN_IN_COMP of ANOTHER_GEN_EX is
   -- Component declaration for NAND_GATE:
   component NAND_GATE
      generic (M: INTEGER);
      port (A: in BIT_VECTOR (M downto 1); Z: out BIT);
   end component;
   -- Component declaration for AND_GATE:
   component AND_GATE
      generic (N: NATURAL := 5);
      port (A: in BIT_VECTOR(1 to N); Z: out BIT);
   end component;
   signal S1, S2, S3, S4: BIT;
   signal SA: BIT_VECTOR (1 to 5);
   signal SB: BIT_VECTOR (2 downto 1);
   signal SC: BIT_VECTOR (1 to 10);
   signal SD: BIT_VECTOR (5 downto 0);
begin
   -- Component instantiations:
   N1: NAND_GATE generic map (6) port map (SD, S1);
   A1: AND_GATE generic map (N => 10) port map (SC, S3);
   A2: AND_GATE port map (SA, S4);
   -- N2: NAND_GATE port map (SB, S2);
end GEN_IN_COMP;
```

For the purposes of this discussion, we shall assume that the components NAND_GATE and AND_GATE are bound to the entities NAND_GATE and AND_GATE described earlier. The component declaration for AND_GATE specifies a value for the generic. When this component is instantiated and a new generic value is assigned using a generic map as in instance A1, the new value, 10, overrides the value specified in the component declaration, 5. When the AND_GATE component is instantiated and no generic map is specified, as in instance A2, the value of the generic specified in the component declaration, 5, is used. In the case of instance N1, again the value supplied by the generic map (i.e., 6) overrides the value assigned to the generic in the entity declaration for NAND_GATE (i.e.,

2). The instance N2, shown as a comment, is illegal, since neither the component instantiation nor the component declaration supply a value for the generic.

The architecture body described earlier uses the more general form of syntax for component declaration and component instantiation statement, which is:

```
-- Component declaration:
component component-name [ is ]
   [ generic ( list-of-generics ) ; ]
   [ port ( list-of-interface-ports ) ; ]
end component [ component-name ] ;

-- Component instantiation statement:
component-label : component-name
            [ generic map ( generic-association-list ) ]
            [ port map ( port-association-list ) ] ;
```

Values for generics may also be specified in a configuration specification or in a configuration declaration. We shall see this later in the section on configurations. A model of a `nor` gate with generic rise and fall delays is shown next.

```
entity NOR2 is
   generic (PT_HL, PT_LH: TIME range 0 ns to TIME'HIGH);
   port (DA, DB: in BIT; DZ: out BIT);
end NOR2;

architecture NOR2_DELAYS of NOR2 is
   signal TEMP: BIT;
begin
   TEMP <= not (DA or DB);
   DZ <= TEMP after PT_HL when TEMP = '0' else
        TEMP after PT_LH;
end NOR2_DELAYS;
```

Since no default values were provided for the generics, the values must be provided when this entity is instantiated or configured.

Consider an `or` gate constructed using two `nor` gates; each `nor` gate has the behavior described previously. The rise and fall delays are specified when the NOR2 component is instantiated. In the following example, different propagation delays are specified in each component instantiation statement.

```
entity OR2 is
   port (A, B: in BIT; C: out BIT);
end OR2;
```

```
architecture OR2_NOR2 of OR2 is
   component NOR2
      generic (PT_HL, PT_LH: TIME);
      port (A, B: in BIT; Z: out BIT);
   end component;
   signal S1: BIT;
begin
   N1: NOR2 generic map (5 ns, 3 ns) port map (A, B, S1);
   N2: NOR2 generic map (6 ns, 5 ns) port map (S1, S1, C);
end;
```

Other uses of generics include modeling ranges of subtypes; for example,

```
subtype ALUBUS is INTEGER range TOP downto 0;
-- TOP is a generic.
```

Generics can also be used to control the number of instantiations of a component in a generate statement (generate statements are discussed in Chapter 10).

7.2 Why Configurations?

A question that is often asked is why are configurations needed? There are two main reasons.

1. Sometimes it may be convenient to specify multiple views for a single entity and use any one of these for simulation. This can be easily done by specifying one architecture body for each view and using a configuration to bind the desired architecture body. For example, there may be three architecture bodies, called FA_BEH, FA_STR, and FA_MIXED, corresponding to an entity FULL_ADDER. Any one of these can be selected for simulation by specifying an appropriate configuration.
2. Similar to the previous case, it may be desirable to associate a component with any one of a set of entities. The component declaration may have its name and the names, types, and number of ports and generics different from those of its entities. For example, a declaration for a component used in a design may be:

```
component OR2
   port (A, B: in BIT; Z: out BIT);
end component;
```

and the entities that the above component may possibly be bound to are:

```
entity OR_GENERIC is
   port (N: out BIT; L, M: in BIT);
end OR_GENERIC;

entity OR_HS is
   port (X,Y: in BIT; Z: out BIT);
end OR_HS;
```

The component names and the entity names, as well as the port names and their order, are different. In one case we may be interested in using the OR_HS entity for the OR2 component, and in another case, the OR_GENERIC entity. This can be achieved by appropriately specifying a configuration for the component. The advantage is that when components are used in a design, arbitrary names for components and their interface ports can be used, and these can later be bound to specific entities prior to simulation.

A configuration is therefore used to bind the following pairs:

1. An architecture body to its entity declaration
2. A component with an entity

Note that a configuration does not have any simulation semantics associated with it; it only specifies how a top-level entity is organized in terms of lower-level entities by specifying the bindings between the entities. The language provides two ways of performing this binding:

1. By using a configuration specification
2. By using a configuration declaration

7.3 Configuration Specification

A configuration specification is used to bind component instantiations to specific entities stored in design libraries. The specification appears in the declarations part of the architecture or block in which the components are instantiated. Binding of a component to an entity can be done on a per-instance basis, for all instantiations of a component, or for a selected set of instantiations of a component. Instantiations of different components can also be bound to the same entity.

Figure 7-1 shows a logic diagram for a one-bit full-adder. Its structural model is described next.

```
library HS_LIB, CMOS_LIB;
entity FULL_ADDER is
  port (A, B, CIN: in BIT; SUM, COUT: out BIT);
end;

architecture FA_STR of FULL_ADDER is
  component XOR2
    port (D1, D2: in BIT; DZ: out BIT);
  end component;
  component AND2
    port (Z: out BIT; A0, A1: in BIT);
  end component;
  component OR2
    port (N1, N2: in BIT; Z: out BIT);
  end component;

  -- Following are four configuration specifications:
  for X1, X2: XOR2
    use entity WORK.XOR2(XOR2BEH);          -- Binding the entity with
        -- more than one instantiation of a component.
  for A3: AND2
    use entity HS_LIB.AND2HS(AND2STR)
      port map (HS_B=>A1, HS_Z=>Z, HS_A=>A0);          -- Binding
        -- the entity with a single instantiation of a component.
  for all: OR2
    use entity CMOS_LIB.OR2CMOS(OR2STR);          -- Binding the
        -- entity with all instantiations of OR2 component.
  for others: AND2
    use entity WORK.A_GATE(A_GATE_BODY)
      port map (A0, A1, Z);          -- Binding the entity with all unbound
                                     -- instantiations of AND2 component.
  signal S1, S2, S3, S4, S5: BIT;
begin
  X1: XOR2 port map (A, B, S1);
  X2: XOR2 port map (S1, CIN, SUM);
  A1: AND2 port map (S2, A, B);
  A2: AND2 port map (S3, B, CIN);
  A3: AND2 port map (S4, A, CIN);
  O1: OR2 port map (S2, S3, S5);
```

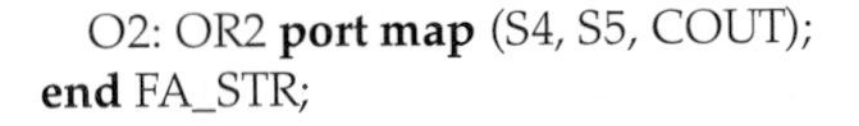

```
   O2: OR2 port map (S4, S5, COUT);
end FA_STR;
```

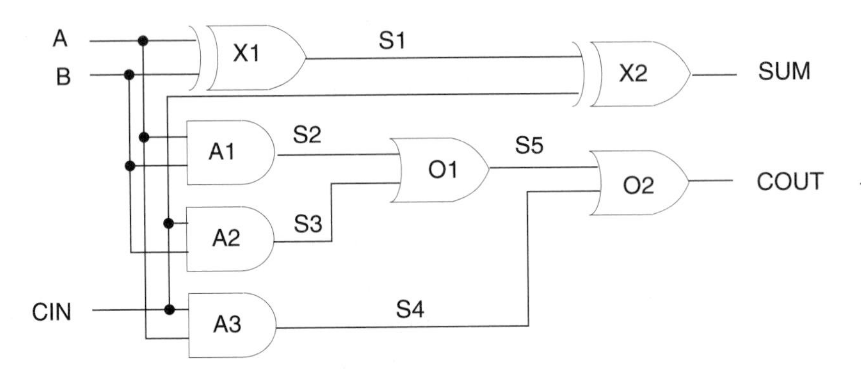

Figure 7-1 A 1-bit full-adder circuit.

There are four configuration specifications appearing in the declarative part of the architecture body. The first specification indicates that instances X1 and X2 of component XOR2 are bound to the entity represented by the entity-architecture pair XOR2 and XOR2BEH, which reside in library WORK. The second specification binds the AND2 component with instantiation label A3 to the entity represented by the entity-architecture pair AND2HS and AND2STR, which is present in design library HS_LIB. The mapping of the component (AND2) ports and the entity (AND2HS) ports is specified using named association; for example, port HS_A of the AND2HS entity is mapped to port A0 of the AND2 component. The third specification implies that all instances of component OR2 use the entity represented by the specified entity-architecture pair present in the design library CMOS_LIB. The last specification implies that all unbound instances of component AND2, that is, instances A1 and A2, are bound to the entity A_GATE using the architecture A_GATE_BODY, which resides in library WORK.

The previous example showed that different instances of the same component can be bound to different entities. Figure 7-2 depicts this binding.

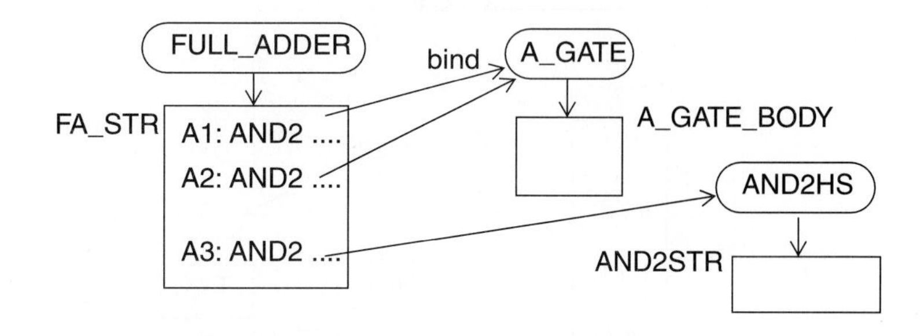

Figure 7-2 Different instances bound to different entities.

Similarly, it is also possible to bind different components to the same entity. An example is shown in Figure 7-3. This figure shows that there is nothing special about the component name AND2. By binding an instance of component AND2 to an entity called OR_GATE, this instance is made to behave as specified in the architecture of entity OR_GATE. Using such a binding may cause confusion to the reader, even though it is correct, and it may be what was intended. Such bindings may sometimes be necessary, for example, while debugging a model we may want to see the effect of specifying an `and` gate to behave like an `or` gate without changing the rest of the description.

This flexibility of being allowed to bind a component instance to any entity may result in a complex maze of bindings. An example is shown in Figure 7-4.

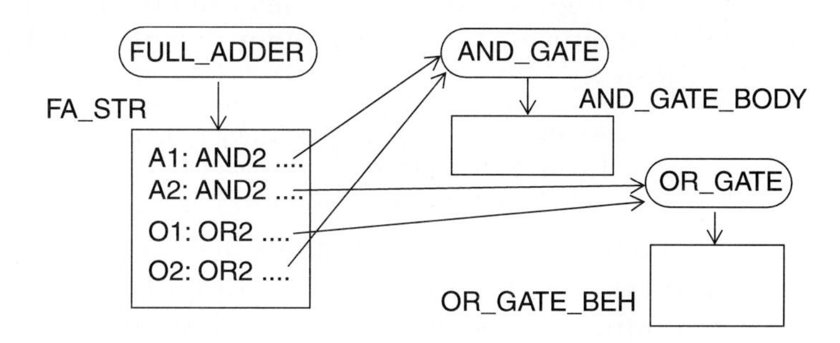

Figure 7-3 Different components bound to same entity.

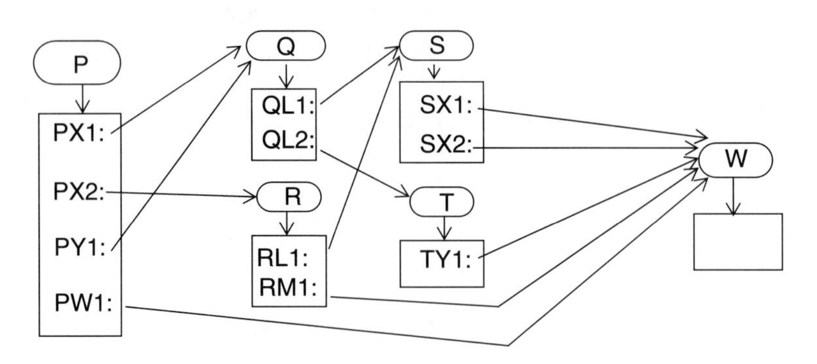

Figure 7-4 A complex maze of bindings.

Entity P has four component instances, PX1 and PX2 of component type PX, PY1 of component type PY, and PW1 of component type PW. Component instances PX1 and PY1 are bound to entity Q, while PX2 is bound to R. Component instances QL1 and QL2 (of component type QL) in entity Q are bound to entities S and T, respectively. Component instances RL1 and RM1 in entity R are bound to entities S and W, respectively. All component instances in S and T, and component PW1 in entity P, are bound to a single entity W. In other words, all the entities P, Q, R, S, and T have been built hierarchically using a single primitive component, W.

The syntax of a configuration specification is:

> **for** *list-of-comp-labels* : *component-name binding-indication* ;

The *binding-indication* specifies the entity represented by the entity-architecture pair, and the generic and port bindings, and one of its forms is:

> **use entity** *entity-name* [(*architecture-name*)]
> [**generic map** (*generic-association-list*)]
> [**port map** (*port-association-list*)] -- Form 1

The list of component labels may be replaced with the keyword **all** to denote all instances of a component; it may also be the keyword **others** to specify all as yet unbound instances of a component. The generic map is used to specify the values for the generics or provide the mapping between the generic parameters of the component and the entity to which it is bound. The port map is used to specify the port bindings between the component and the bound entity. Additional examples of configuration specifications appear in the following architecture body.

```
architecture DUMMY of DUMMY is
  component NOR_GATE
    generic (RISE_TIME, FALL_TIME: TIME);
    port (S0, S1: in BIT; Q: out BIT);
  end component;
  component AND2_GATE
    port (DOUT: out BIT; DIN: in BIT_VECTOR);
  end component;
  for N1, N2: NOR_GATE
    use entity WORK.NOR2(NOR2_DELAYS)
      generic map ( PT_HL => FALL_TIME,
                    PT_LH => RISE_TIME)
      port map (S0, S1, Q);
  for all: AND2_GATE
    use entity WORK.AND2 (GENERIC_EX)
      generic map (10)
      port map (A => DIN, Z => DOUT);
  signal WR, RD, RW, S1, S2: BIT;
  signal SA: BIT_VECTOR (1 to 10);
begin
  N1: NOR_GATE generic map (2 ns, 3 ns) port map (WR, RD, RW);
  A1: AND2_GATE port map (S1, SA);
  N2: NOR_GATE generic map (4 ns, 6 ns)
                port map (S1, SA(2), S3);
end DUMMY;
```

The entity declarations for the entities that are bound to components NOR_GATE and AND2_GATE are:

```
entity NOR2 is
  generic (PT_HL, PT_LH: TIME);
  port (DA, DB: in BIT; DZ: out BIT);
end NOR2;

entity AND2 is
  generic (N: NATURAL := 5);
  port (A: in BIT_VECTOR(1 to N); Z: out BIT);
end AND2;
```

In the binding for N1 and N2, the generic map specifies the mapping of generic names from the entity NOR2 to the component NOR_GATE using named association. The generic values supplied in the instantiations are therefore passed to the NOR2 entity through this mapping. The port binding is specified using positional association; that is, ports S0, S1, and Q of NOR_GATE component map to ports DA, DB, and DZ, respectively, of the NOR2 entity. The configuration specification for the AND2_GATE specifies the value 10 for the generic explicitly (using positional association), which overrides the default value 5 specified in the entity declaration for the AND2 entity. The component declaration for AND2_GATE should not specify any generics, since the values are passed directly to the actual generics of the entity. The port mapping for the AND2_GATE is specified using named association.

How are the generic map and port map values in a component instantiation passed into its bound entity via the configuration specification? A look at the elaboration of a component instantiation helps us to understand this. Let us take the N1 instantiation in the previous architecture body as an example. Elaboration transforms this component instantiation into the following block statement.

```
N1: block                        -- A block for the component instantiation.
  -- Generics of the component:
  generic (RISE_TIME, FALL_TIME: TIME);
  -- Generic map in instantiation:
  generic map (RISE_TIME => 2 ns, FALL_TIME => 3 ns);
  port (S0, S1: in BIT; Q: out BIT);          -- Ports of component.
  port map (S0 => WR, S1 => RD, Q => RW);             -- Port map in
                                                      -- instantiation.
begin
  NOR2: block                            -- A block for the bound entity.
    generic (PT_HL, PT_LH: TIME);                     -- Its generics.
    generic map (PT_HL => FALL_TIME, PT_LH => RISE_TIME);
      -- Generic map in configuration specification.
    port (DA, DB: in BIT; DZ: out BIT);               -- Its ports.
    port map (DA => S0, DB => S1, DZ => Q);           -- Port map in
                                                      -- specification.
    -- Other declarations in the architecture
    -- body NOR2_DELAYS appear here.
  begin
    -- Statements in architecture body NOR2_DELAYS appear here.
  end block NOR2;
end block N1;
```

The block N1 is created from the component instantiation of NOR_GATE. The generic map and port map for this block are the generic map and port map specified in the component instantiation for N1; that is, they specify the mapping between the values and signals in the component instantiation statement with the generics and ports of component NOR_GATE. The inner block with label NOR2 represents the entity NOR2 that the component instantiation N1 is bound to in the configuration specification. The generic map and port map of this block specify the generic map and port map that appear in the configuration specification; that is, they specify the mapping between the component NOR_GATE and the entity NOR2.

7.4 Configuration Declaration

Configuration specifications have to appear in an architecture body. Therefore, to change a binding it is necessary to change the architecture body and re-analyze it. This may be cumbersome and time-consuming. To avoid this, a configuration declaration may be used to specify a binding.

A configuration declaration is a separate design unit. It therefore allows for late binding of components, that is, the bindings can be performed after the architecture body has been written. It is also possible to have more than one configuration declaration for an entity, each of which defines a different set of bindings for components in a single architecture body, or possibly specifies a unique entity-architecture pair.

The typical format of a configuration declaration is:

```
configuration configuration-name of entity-name is
  block-configuration
end [ configuration ] [ configuration-name ] ;
```

It declares a configuration with the name *configuration-name*, which applies to the entity *entity-name*. A *block-configuration* defines the binding of components in a block, where a block may be an architecture body, a block statement, or a generate statement. Bindings of components defined in a block statement and in a generate statement are discussed in Chapter 10. A block configuration is a recursive structure of the form:

```
for block-name
  component-configurations
  block-configurations
end for;
```

The *block-name* is the name of an architecture body, a block statement label, or a generate statement label. The top-level block is always an architecture body. A *component-configuration* binds components that appear in a block to entities, and is of the form:

```
for list-of-comp-labels : comp-name [ binding-indication ; ]
  [ block-configuration ]
end for;
```

The block configuration that appears within a component configuration defines the bindings of components at the next level of hierarchy in the entity-architecture pair specified by the binding indication.

There are two other forms of binding indication in addition to the one shown in the previous section. These are:

```
use configuration configuration-name          -- Form 2
use open                                      -- Form 3
```

In form 2, the binding indication specifies that the component instances are to be bound to a configuration of a lower-level entity as specified by the configuration name. This implies that a configuration declaration with such a name must exist. In form 3, the binding indication indicates that the binding is not yet specified and is to be deferred. Both these forms of binding indication may also be used in a configuration specification.

Here is an example of a configuration declaration that specifies the component configurations for all component instances in architecture FA_STR of entity FULL_ADDER described in the previous section.

```
library CMOS_LIB;
configuration FA_CON of FULL_ADDER is
  for FA_STR
    use WORK.all;
    for A1, A2, A3: AND2
      use entity CMOS_LIB.BIGAND2 (AND2STR);
    end for;
    for others: OR2           -- use defaults, i.e. use OR2 from
                              -- library WORK.
    end for;
    for all: XOR2
      use configuration WORK.XOR2CON;
    end for;
  end for;
end FA_CON;
```

The configuration with name FA_CON binds architecture FA_STR with the FULL_ADDER entity. The components within this architecture body, instances A1, A2, and A3, are bound to the entity BIGAND2 that exists in the design library CMOS_LIB. For all instances of component OR2, the default bindings are used; these are the entities in the working library with the same names as the component names. The last component configuration shows a different type of binding indication. In this case, all component instances are bound to a configuration instead of an entity-architecture pair. All instances of component XOR2 are bound to a configuration with name XOR2CON, which exists in the working library. This type of binding may also be specified in a configuration specification.

The power of the configuration declaration lies in the fact that the sub-components in an entire hierarchy of a design can be bound using a single configuration declaration. For example, consider a full-adder circuit composed of two half-adders and an `or` gate. The half-adder circuit is in turn composed of `xor` and `and` gates. The hierarchy for this full-adder is shown in Figure 7-5.

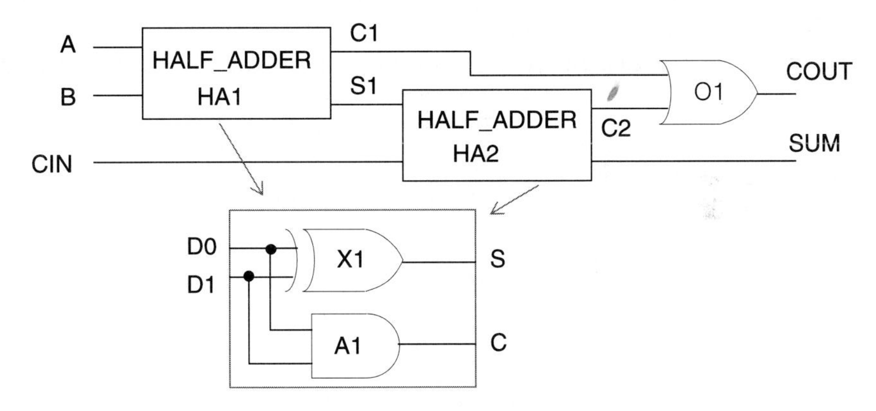

Figure 7-5 A hierarchical 1-bit full-adder.

The structural models for the full-adder and half-adder circuits are described next. A configuration declaration that specifies the bindings for components used in the entire hierarchy of the full-adder is also shown.

```
entity FULL_ADDER is
  port (A, B, CIN: in BIT; COUT, SUM: out BIT);
end FULL_ADDER;
```

```
architecture FA_WITH_HA of FULL_ADDER is
  component HALF_ADDER
    port (HA, HB: in BIT; HS, HC: out BIT);
  end component;
  component OR2
    port (A, B: in BIT; Z: out BIT);
  end component;
  signal S1, C1, C2: BIT;
begin
  HA1: HALF_ADDER port map (A, B, S1, C1);
  HA2: HALF_ADDER port map (S1, CIN, SUM, C2);
  O1: OR2 port map (C1, C2, COUT);
end FA_WITH_HA;

entity HA is
  port (D0, D1: in BIT; S, C: out BIT);
end HA;

architecture HA_STR of HA is
  component XOR2
    port (X,Y: in BIT; Z: out BIT);
  end component;
  component AND2
    port (L, M: in BIT; N: out BIT);
  end component;
begin
  X1: XOR2 port map (D0, D1, S);
  A1: AND2 port map (D0, D1, C);
end HA_STR;

-- A configuration for the FULL_ADDER that is built
-- using half-adders.
library ECL;
configuration FA_HA_CON of FULL_ADDER is
  for FA_WITH_HA                          -- Top-level block configuration.
    for HA1, HA2: HALF_ADDER
      use entity WORK.HA(HA_STR)
        port map (D0 => HA, D1 => HB, S => HS, C => HC);
      for HA_STR                          -- Nested block configuration.
        for all: XOR2
          use entity WORK.XOR2(XOR2);
        end for;
        for A1: AND2
          use configuration ECL.AND2CON;
        end for;
      end for;
    end for;
```

```
    for O1: OR2
      use configuration WORK.OR2CON;
    end for;
  end for;
end FA_HA_CON;
```

The top-level block configuration specifies the bindings of component instances present in the architecture body FA_WITH_HA. Instances HA1 and HA2 are bound to an entity specified by the entity-architecture pair, entity HA and architecture HA_STR. The nested block configuration specifies the binding of component instances present in the architecture body HA_STR. In this way, a configuration can be nested to any arbitrary depth and can be used to bind all components in a hierarchy.

The previous example shows that, when components in a hierarchy are bound, a single configuration declaration may be used to replace a set of configuration specifications. If configuration specifications were used in the earlier example, they would have to be included separately in the architecture bodies FA_WITH_HA and HA_STR, and then these bodies would have to be recompiled every time a binding is changed. Note that a component instance must not be bound in both a configuration specification and a configuration declaration.

Therefore, configurations provide the mechanism by which architecture bodies may contain technology-independent components. Technology-specific mappings can be specified separately using configuration declarations.

7.5 Default Rules

A large amount of overhead is introduced if the bindings for every component in a design must be individually specified. Fortunately, the language provides default binding rules. For each unbound component instance:

- An entity that is visible and has the same name as that of the component is used to bind to the instance; if no such entity exists, a default binding indication of "**use open**" is used.

- The most recently analyzed architecture for that entity is used; it is an error if no architecture body exists.
- For each port or generic in the component instance, there must exist a corresponding port or generic in the entity that matches by name, type, and mode; any ports or generics in the entity that are unassociated are treated as `open`; if any match fails, it is an error.

These default rules help to avoid writing specific bindings in cases where the component names are the same as the entity names. Another advantage of these default rules is that it allows for the usage of standard component names such as SN7400 and SN7402. In cases such as these, no bindings are necessary. However, a minimum configuration may sometimes be necessary. This could be of the form:

```
library TTL_LIB;
configuration TLC_CON of TLC is
  for TLC_STRUCTURE
    use TTL_LIB.all;
  end for;
end TLC_CON;
```

TLC is an entity that has some component instantiations defined within its architecture body. The `use` clause "**use** TTL_LIB.**all**" makes all entities in library TTL_LIB visible. Therefore, these entities will get bound to the component instances in the architecture body by virtue of the default rules.

7.6 Conversion Functions

It is possible for the port types of an entity to differ from the port types of the component being bound to. For example, given the component declaration:

```
component COUNTER
  port (CTR: inout BIT_VECTOR;
        CLK, RST: in BIT;
        PAR: out BIT);
end component;
```

the component COUNTER is bound to an entity with the declaration:

```
entity COUNTER is
   port (Q: inout STD_LOGIC_VECTOR(3 downto 0);
         CLOCK, RESET: in STD_LOGIC;
         PARITY: out STD_LOGIC);
end COUNTER;
```

This scenario occurs quite often in practice, for example, when a vendor supplies a library of entity declarations using the vendor's favorite type, while the user models a design using a different data type. In such a case, *conversion functions* can be used to convert the values from one type to another by specifying this function in the association list of the binding.

In the binding information of a configuration specification, when data flows from a component to an entity, that is, for an input port, the conversion function is specified with the component port name; this means that the conversion function is applied on the value of the component port before assigning it to the entity port. For an output port, the conversion function is specified with the entity port name, since data flows from entity to component, while for an inout port, the conversion function is specified on both the component port and the entity port.

Here is an example.

```
for all: COUNTER use entity WORK.COUNTER
   port map (
      TO_BITVECTOR(Q) => TO_STDLOGICVECTOR(CTR),
      CLOCK => TO_STDLOGIC(CLK),
      RESET => TO_STDLOGIC(RST),
      TO_BIT(PARITY) => PAR);
```

CLOCK and RESET are input ports of the entity. Therefore, data flows from the component ports (which are CLK and RST) to the entity ports. In this case, the conversion function TO_STDLOGIC appears on the component ports. The value of the component port is the parameter to the conversion function. The value returned by the conversion function is then passed to the entity port.

For the output port PARITY, the conversion function is specified with the entity port. The entity port value is passed as a parameter to the conversion function TO_BIT, and its return value is passed on to the component port PAR.

For the inout port Q, conversion functions appear on both the component ports and the entity ports, since data travels both ways. When data is passing from component to entity, the conversion function TO_STDLOGICVECTOR is called, and the return value is passed to the entity port. If data is flowing from the entity to the component, the conversion function TO_BITVECTOR is called, and its return value is assigned to the component port CTR.

Conversion functions can be used wherever there is an association of generics, ports, or parameters, such as in parameter associations in a function call or a procedure call, in port map and generic map of a block statement, or a component instantiation statement.

7.7 Direct Instantiation

A component declaration declares the interface of a component. This component can then be instantiated using a component instantiation statement. However, before the entity that contains the component can be simulated, the component instance needs to be bound or linked to an entity-architecture pair or to a configuration. This binding is specified using additional constructs, that is, by using a configuration specification or a configuration declaration.

However, it is possible to directly instantiate the entity-architecture pair or a configuration in a component instantiation statement. This saves the additional binding step necessary when using components. Here are the two additional forms of the component instantiation statement that can be used to directly instantiate an entity or a configuration.

component-label : **entity** *entity-name* [(*architecture-name*)]
    [**generic map** (*generic-association-list*)]
    [**port map** (*port-association-list*)] ;

component-label : **configuration** *configuration-name*
    [**generic map** (*generic-association-list*)]
    [**port map** (*port-association-list*)] ;

Here is the description of the hierarchical 1-bit full-adder using direct instantiation. Note that no configuration declaration is necessary or possible in this case, since the component instantiations directly instantiate the appropriate entity-architecture pairs or configurations. Also, no component declarations are necessary or possible. The description is therefore much more compact.

```
entity FULL_ADDER is
   port (A, B, CIN: in BIT; COUT, SUM: out BIT);
end FULL_ADDER;

architecture FA_WITH_HA of FULL_ADDER is
   signal S1, C1, C2: BIT;
begin
   HA1: entity WORK.HA (HA_STR) port map (A, B, S1, C1);
   HA2: entity WORK.HA (HA_STR) port map (S1, CIN, SUM, C2);
   O1: configuration WORK.OR2CON port map (C1, C2, COUT);
end FA_WITH_HA;

entity HA is
   port (D0, D1: in BIT; S, C: out BIT);
end HA;

library ECL;
architecture HA_STR of HA is
begin
   X1: entity WORK.XOR2(XOR2) port map (D0, D1, S);
   A1: configuration ECL.AND2CON port map (D0, D1, C);
end HA_STR;
```

Here is another example. This is the full-adder example with architecture FA_STR, described earlier in Section 7.3. Note that, again in this case, no configuration specifications and component declarations are necessary, thereby making the description very compact.

```
library HS_LIB, CMOS_LIB;
entity FULL_ADDER is
   port (A, B, CIN: in BIT; SUM, COUT: out BIT);
end FULL_ADDER;

architecture FA_STR of FULL_ADDER is
   signal S1, S2, S3, S4, S5: BIT;
begin
   X1: entity WORK.XOR2 (XOR2BEH) port map (A, B, S1);
   X2: entity WORK.XOR2 (XOR2BEH) port map (S1, CIN, SUM);
   A1: entity WORK.A_GATE (A_GATE_BODY) port map (S2, A, B);
   A2: entity WORK.A_GATE (A_GATE_BODY)
         port map (S3, B, CIN);
   A3: entity HS_LIB.AND2HS (AND2STR) port map (S4, A, CIN);
```

```
    O1: entity CMOS_LIB.OR2CMOS (OR2STR)
        port map (S2, S3, S5);
    O2: entity CMOS_LIB.OR2CMOS (OR2STR)
        port map (S4, S5, COUT);
end FA_STR;
```

7.8 Incremental Binding

It is possible to delay the bindings of ports and generics and to override pre-specified generics using the incremental binding mechanism. Bindings present in a configuration specification represent the primary bindings. However, this may not contain the complete binding information. Missing information could be any of the following:

- Unassociated ports and generics
- Open ports and generics

Here is an example that contains incomplete binding information.

```
entity FULL_ADDER is
  port (A, B, CIN: in BIT; SUM, COUT: out BIT);
end FULL_ADDER;

architecture FA_STR_INCR of FULL_ADDER is
  component XOR2
    port (D1, D2: in BIT; DZ: out BIT);
  end component;
  component AND2
    port (Z: out BIT; A0, A1: in BIT);
  end component;
  component OR2
    port (N1, N2: in BIT; Z: out BIT);
  end component;

  for X1, X2: XOR2
    use entity WORK.MY_XOR2;          -- No generic map and port map
      -- specified. Since architecture name is not specified,
      -- default is the most recently analyzed architecture.
  for others: AND2 use entity WORK.MY_AND2(ARCH_BODY)
    port map (HS_B => A1, HS_A => open);          -- Port HS_Z is not
      -- associated and port HS_A is open.
  for all: OR2 use entity WORK.MY_OR2
```

```
        generic map (TPHL => 2 ns, TPLH => 3 ns);     -- These generic
          -- values will later be overridden by the ones specified in the
          -- configuration declaration.
begin
  . . .
end FA_STR_INCR;
```

The three entities to which the above components are bound are:

```
entity MY_XOR2 is
  generic (TPHL, TPLH: TIME);
  port (XA, XB: in BIT; XZ: out BIT);
end MY_XOR2;

entity MY_AND2 is
  port (HS_A, HS_B: in BIT; HS_Z: out BIT);
end MY_AND2;

entity MY_OR2 is
  generic (TPHL, TPLH: TIME);
  port (N1, N2: in BIT; Z: out BIT);
end MY_OR2;
```

In the bindings for the XOR2 components, the port and generic maps are missing. In the AND2 component bindings, port HS_Z is not associated, and in the bindings for the OR2 component, generic values are specified. However, generic values specified in a configuration declaration can override the generic values specified in a configuration specification. All the extra and missing information can later be supplied in a configuration declaration, which may exist in a different file. Here is a configuration declaration with the missing information.

```
configuration FA_INCREMENTAL of FULL_ADDER is
  for FA_STR_INCR
    -- First component configuration:
    for X1, X2: XOR2
      port map (A0, A1, Z)
      generic map (TPHL => 2 ns, TPLH => 5 ns);
    end for;
    -- Second component configuration:
    for all: AND2
      port map (HS_A => '1', HS_Z => Z);
    end for;
    -- Third component configuration:
    for all: OR2
      generic map (TPHL => 4 ns, TPLH => 6 ns);
    end for;
  end for;
end FA_INCREMENTAL;
```

In the configuration declaration, there is no need to specify the entity name to which the components are bound (by using "**use entity** . . ."), since this information is already provided in the configuration specifications that are present inside the architecture body. In the first component configuration, the missing port map and generic map is provided. In the second component configuration, the unassociated port and the open port are bound. In the third component configuration, the new set of generic values specified override the generic values "TPHL => 2 ns, TPLH => 3 ns" specified in the configuration specification.

❑

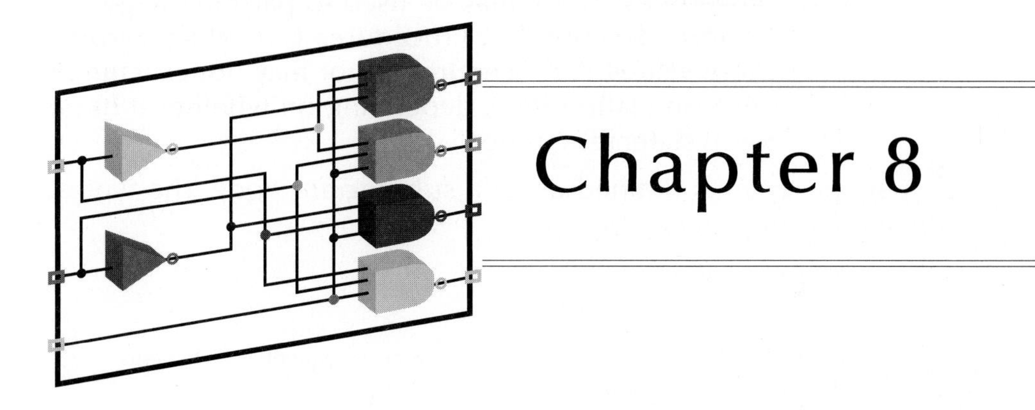

Chapter 8

Subprograms and Overloading

This chapter describes the two kinds of subprograms: procedures and functions. A function call can be used within an expression. A procedure call may be a sequential or concurrent statement. It is possible for two or more subprograms to have the same name. This is called overloading. This chapter also presents this important concept, and explains operator overloading as well.

8.1 Subprograms

A subprogram defines a sequential algorithm that performs a certain computation. There are two kinds of subprograms:

1. *Functions*: These are usually used for computing a single value. A function executes in zero simulation time.

2. *Procedures*: These may be used to partition large behavioral descriptions. Procedures can return zero or more values. A procedure may or may not execute in zero simulation time, depending on whether it has a `wait` statement or not.

A subprogram is defined using a *subprogram body*. The typical format for a subprogram body is:

subprogram-specification **is**
  subprogram-item-declarations
begin
  subprogram-statements -- Same as sequential-statements.
end [**function** | **procedure**] [*subprogram-name*] ;

The *subprogram-specification* specifies the name of a subprogram and defines its interface, that is, it defines the *formal parameter* names, their class (i.e., **signal**, **variable**, **file**, or **constant**), their type, and their mode (whether they are **in**, **out**, or **inout**). Parameters of mode **in** are read-only parameters; these cannot be updated within a subprogram body. Parameters of mode **out** are write-only parameters; their values cannot be used but can only be updated within a subprogram body. Parameters of mode **inout** can be read as well as updated. Files have no mode; they can be read or written to, depending on how the files were opened.

Actuals in a subprogram call are used to pass values to and from a subprogram. Only an actual signal may be associated with a formal parameter of the **signal** class. Only an actual variable may be associated with a parameter of the **variable** class. Only a file may be an actual for a file parameter. An expression may be used to pass a value to a parameter of **constant** class.

When parameters are of a **variable** or **constant** class, values are passed to the subprogram by value. Arrays may or may not be passed by reference. Files are passed by reference. For signals, the reference to the signal, its driver, or both are passed into the subprogram. What this means is that any assignment to a signal in a procedure (signals cannot be assigned values in a function because the parameters are restricted to be of mode **in**) affects the actual signal's driver immediately and is independent of whether the procedure terminates or not. For a signal of any mode, none of the signal-valued attributes, 'STABLE, 'QUIET, 'DELAYED, or 'TRANSACTION (attributes are discussed in Chapter 10), can be used in a subprogram body.

The type of an actual in a subprogram call must match its corresponding formal parameter. If the formal parameter belongs to an unconstrained type, the size of this parameter is determined from the actual that is passed in.

The *subprogram-item-declarations* part contains a set of declarations (e.g., type and object declarations) that are accessible for use within the subprogram. These declarations come into effect every time the subprogram is called. Variables are created and initialized every time the subprogram is called. They remain in existence until the subprogram completes. This is in contrast with variables in a process statement, which get initialized only once at the start of simulation and persist throughout the entire simulation.

The *subprogram-statements* part contains sequential statements that define the computation to be performed by the subprogram. A return statement is a special statement allowed only within subprograms. The format of a return statement is:

return [*expression*] ;

The return statement causes the subprogram to terminate and control to be returned to the calling program. All functions must have a return statement with an expression; the value of the expression in the return statement is returned to the calling program. For procedures, objects of mode **out** and **inout** return their values to the calling program. Procedures may also affect signals and variables declared globally to it.

The *subprogram-name* appearing at the end of a subprogram body, if present, must be the same as the function or procedure name specified in the subprogram specification part.

8.1.1 Functions

Functions are used to describe frequently used sequential algorithms that return a single value. This value is returned to the calling program using a return statement. Some of their common uses are as resolution functions and type conversion functions. The following is an example of a function body.

```
function LARGEST (TOTAL_NO: INTEGER; SET: PATTERN)
    return REAL is
  -- PATTERN is elsewhere defined to be a type of 1-D array of
  -- non-negative floating-point values.
  variable RETURN_VALUE: REAL := 0.0;
begin
  for K in SET'RANGE loop
    if SET(K) > RETURN_VALUE then
      RETURN_VALUE := SET(K);
    end if;
  end loop;
  return RETURN_VALUE;
end LARGEST;
```

Variable RETURN_VALUE comes into existence with an initial value of 0.0 every time the function is called. It ceases to exist after the function terminates.

The general syntax of a subprogram specification for a function body is:

```
[ pure | impure ] function function-name ( parameter-list )
  return return-type
```

A *pure function* is one that returns the same value each time the function is called with the same set of actuals. An *impure function* is one that potentially returns different values each time it is called with the same set of actuals. For example, NOW is an impure function because it returns different values when called at different times. If neither keyword **pure** nor **impure** is present in the function specification, the function is, by default, a pure function.

The *parameter-list* describes the list of formal parameters for the function. The only mode allowed for the parameters is mode **in**. Also, only constants and signal objects can be passed in as parameters. The default object class is **constant**. For example, in function LARGEST, TOTAL_NO is a constant, and its value cannot be modified within the function body. Another example of a function body is shown next. This function returns true if a rising edge has been detected on the input signal. The function has been explicitly marked as a pure function.

```
pure function VRISE (signal CLOCK_NAME: BIT)
    return BOOLEAN is
begin
  return CLOCK_NAME = '1' and CLOCK_NAME'EVENT;
end VRISE;
```

Here are two examples of impure functions.

```
impure function RANDOM (SEED: REAL) return REAL is
  variable NUM: REAL;
  attribute FOREIGN of RANDOM: function is "NUM = rand(seed)";
begin
  return NUM;
end RANDOM;

impure function USED (TO_ALLOCATE: POSITIVE)
    return POSITIVE is
  -- ALLOCATED is a shared variable declared elsewhere, but
  -- visible to this function.
begin
  ALLOCATED := ALLOCATED + TO_ALLOCATE;
  return ALLOCATED;
end function USED;          -- Keyword function after end is optional.
```

A function call is an expression and can also be used in larger expressions. For example,

```
SUM := SUM + LARGEST(MAX_COINS, COLLECTION);
```

A function call has the form:

function-name (*list-of-actuals*)

The actuals may be associated by position (the first actual corresponds to the first formal parameter, the second actual corresponds to the second parameter, and so on) or using named association (the association of actuals and formal parameters are explicitly specified). The function call in the last example used positional association. An equivalent function call using named association is:

```
LARGEST (SET => COLLECTION, TOTAL_NO => MAX_COINS)
```

Functions are frequently used to perform type conversion. Here is an example of a function that converts a value from the STD_ULOGIC type to a value from the CHARACTER type.

```
function TO_CHARACTER (ARG: STD_ULOGIC)
    return CHARACTER is
begin
  case ARG is
    when 'U'    => return 'U';
    when 'X'    => return 'X';
    when '0'    => return '0';
    when '1'    => return '1';
    when 'Z'    => return 'Z';
    when 'W'    => return 'W';
```

```
      when 'L'     => return 'L';
      when 'H'     => return 'H';
      when '-'     => return '-';
    end case;
  end TO_CHARACTER;
```

A function call of the form:

```
TO_CHARACTER (STD_ULOGIC'('U'))
```

returns the character 'U'. However, there is a much more efficient way of performing type conversion, by using look-up tables. Here is the TO_CHARACTER function described using look-up tables.

```
type LOOK_UP is array (STD_ULOGIC) of CHARACTER;
constant TO_CHARACTER: LOOK_UP :=
    ('U' => 'U', 'X' => 'X', '0' => '0',
     '1' => '1', 'Z' => 'Z', 'W' => 'W', 'L' => 'L', 'H' => 'H', '-' => '-');
```

Then the following expression,

```
TO_CHARACTER (STD_ULOGIC'('U'))
```

gives the character 'U'. Note that in both schemes, the calling line to perform type conversion does not change. However, in the look-up table scheme, access time is simply the time to look up a table, while in the function scheme, the overhead is that of a function call, a case statement, and a return statement, which is substantially more.

Here is an example of a function that performs byte reversal. The lower-order bits are swapped with the higher-order bits, for example, for a 8-bit vector, the function swaps the 7th bit with the 0th bit, the 6th bit with the 1st bit, and so on.

```
function BYTE_REVERSAL (ARG: STD_LOGIC_VECTOR)
    return STD_LOGIC_VECTOR is
  variable RESULT:
      STD_LOGIC_VECTOR(ARG'REVERSE_RANGE);
begin
  for J in ARG'RANGE loop
    RESULT(J) := ARG(J);
  end loop;
  return RESULT;
end BYTE_REVERSAL;
```

A function call:

```
BYTE_REVERSAL ("11001110")
```

returns the value:

```
"01110011"
```

8.1.2 Procedures

Procedures allow decomposition of large behaviors into modular sections. In contrast to a function, a procedure can return zero or more values using parameters of mode **out** and **inout**. The syntax for the subprogram specification for a procedure body is:

procedure *procedure-name* (*parameter-list*)

The *parameter-list* specifies the list of formal parameters for the procedure. Parameters may be constants, variables, or signals, and their modes may be **in**, **out**, or **inout**. If the class of a parameter is not explicitly specified, the class is by default a **constant** if the parameter is of mode **in**; otherwise, it is a variable if the parameter is of mode **out** or **inout**.

A simple example of a procedure body is shown next. It describes the behavior of an arithmetic logic unit.

```
type OP_CODE is (ADD, SUB, MUL, DIV, LT, LE, EQ);
. . .
procedure ARITH_UNIT (A, B: in INTEGER; OP: in OP_CODE;
                Z: out INTEGER; ZCOMP: out BOOLEAN) is
begin
  case OP is
    when ADD => Z := A + B;
    when SUB => Z := A – B;
    when MUL => Z := A * B;
    when DIV => Z := A / B;
    when LT => ZCOMP := A < B;
    when LE => ZCOMP := A <= B;
    when EQ => ZCOMP := A = B;
  end case;
end ARITH_UNIT;
```

The following is another example of a procedure body. This procedure rotates the specified signal vector ARRAY_NAME, starting from bit START_BIT to bit STOP_BIT, by the ROTATE_BY value. The object class for parameter ARRAY_NAME is explicitly specified. The variable FILL_VALUE is automatically initialized to '0' every time the procedure is called.

```
procedure ROTATE_LEFT
    (signal ARRAY_NAME: inout BIT_VECTOR;
    START_BIT, STOP_BIT: in NATURAL;
    ROTATE_BY: in POSITIVE) is
  variable FILL_VALUE: BIT;        -- Every time the procedure is called,
```

```
                        -- initial value of FILL_VALUE is BIT'LEFT, which is '0'.
begin
  assert STOP_BIT > START_BIT
    report "STOP_BIT is not greater than START_BIT."
    severity NOTE;
  for MACVAR3 in 1 to ROTATE_BY loop
    FILL_VALUE := ARRAY_NAME(STOP_BIT);
    for MACVAR1 in STOP_BIT downto (START_BIT+1) loop
      ARRAY_NAME(MACVAR1) <=
                ARRAY_NAME(MACVAR1–1);
    end loop;
    ARRAY_NAME(START_BIT) <= FILL_VALUE;
  end loop;
end procedure ROTATE_LEFT;          -- Keyword procedure after end is
                                    -- optional.
```

Procedures are invoked by using procedure calls. A procedure call can be either a sequential statement or a concurrent statement; this is based on where the actual procedure call statement is present. If the call is inside a process statement or another subprogram, it is a sequential procedure call statement; otherwise, it is a concurrent procedure call statement. The syntax of a procedure call statement is:

[*label* :] *procedure-name* (*list-of-actuals*) ;

The actuals specify the expressions, signals, variables, or files, that are to be passed into the procedure and the names of objects that are to receive the computed values from the procedure. Actuals may be specified using positional association or named association. For example,

```
ARITH_UNIT (D1, D2, ADD, SUM, COMP);          -- Positional association.
ARITH_UNIT (Z => SUM, B => D2, A => D1,
        OP => ADD, ZCOMP => COMP);            -- Named association.
```

A sequential procedure call statement is executed sequentially with respect to the sequential statements surrounding it. A concurrent procedure call statement is executed whenever an event occurs on one of the parameters that is a signal of mode **in** or **inout**. Semantically, a concurrent procedure call is equivalent to a process with a sequential procedure call and a `wait` statement that waits for an event on the signal parameters of mode **in** or **inout**. Here is an example of a concurrent procedure call and its equivalent process statement.

```
architecture DUMMY_ARCH of DUMMY is
  -- Following is a procedure body:
```

```
    procedure INT_2_VEC (signal D: out BIT_VECTOR;
      START_BIT, STOP_BIT: in NATURAL;
      signal VALUE: in INTEGER) is
    begin
      -- Procedure behavior here.
    end INT_2_VEC;
begin
  -- This is an example of a concurrent procedure call:
  INT_2_VEC (D_ARRAY, START, STOP, SIGNAL_VALUE);
end DUMMY_ARCH;

-- The equivalent process statement for the concurrent
-- procedure call is:
  process
  begin
    INT_2_VEC (D_ARRAY, START, STOP, SIGNAL_VALUE);
    -- This is now a sequential procedure call.
    wait on SIGNAL_VALUE;
    -- Since SIGNAL_VALUE is an input signal.
  end process;
```

A procedure can be used as either a concurrent statement or a sequential statement. Concurrent procedure calls are useful in representing frequently used processes.

Concurrent procedures can also be marked as postponed, such as:

```
postponed procedure INT_2_VEC (signal D: out BIT_VECTOR;
    START_BIT, STOP_BIT: in NATURAL;
    signal VALUE: in INTEGER) is
begin
  -- Procedure behavior here.
end INT_2_VEC;
```

The semantics of a postponed concurrent procedure call is equivalent to the semantics of its equivalent process statement, which is a postponed process.

A procedure body can have a `wait` statement, while a function cannot. Functions are used to compute values that are available instantaneously. Therefore, a function cannot be made to wait; for example, it cannot call a procedure with a `wait` statement in it. A process that calls a procedure with a `wait` statement cannot have a sensitivity list. This follows from the fact that a process cannot be sensitive to signals and also be made to wait simultaneously. Since a procedure can have a `wait` statement, any variables and constants declared in the procedure retain their values through this wait period and cease to exist only when the procedure terminates.

8.1.3 Declarations

A subprogram body may appear in the declarative part of the block in which a call is made. This is not convenient if the subprogram is to be shared by many entities. In such cases, the subprogram body may be described possibly in a package body and then, in the package declaration, the corresponding *subprogram declaration* is specified. If this package declaration is made visible in another design unit using an `use` clause, the subprogram can be used in this design unit. A subprogram declaration describes the subprogram name and the list of parameters without describing the internal behavior of the subprogram, that is, it describes the interface for the subprogram. The syntax of a subprogram declaration is:

subprogram-specification ;

Four examples of procedure and function declarations are shown next.

```
procedure ARITH_UNIT (A, B: in INTEGER; OP: in OP_CODE;
                                  Z: out INTEGER; ZCOMP: out BOOLEAN);
pure function VRISE (signal CLOCK_NAME: BIT)
    return BOOLEAN;
impure function RANDOM (SEED: REAL) return REAL;
function LARGEST (TOTAL_NO: INTEGER; SET: PATTERN)
    return REAL;                                  -- This is a pure function.
```

Another reason subprogram declarations are necessary is to allow two subprograms to call each other recursively; for example,

```
procedure PROC_P (. . .) . . .
  variable LOCAL: . . .
begin
  . . .
  LOCAL := FUNC_Q (B);                              -- illegal function call.
  . . .
end PROC_P;

function FUNC_Q (. . .) . . .
begin
  . . .
  PROC_P(. . .);
  . . .
end FUNC_Q;
```

The call to function FUNC_Q in procedure PROC_P is illegal, since FUNC_Q has not yet been declared. This can be corrected by writing the function declaration for FUNC_Q before procedure PROC_P.

8.2 Subprogram Overloading

Sometimes it is convenient to have two or more subprograms with the same name. In such a case, the subprogram name is said to be *overloaded*; also, the corresponding subprograms are said to be overloaded. For example, consider the following two declarations.

```
function COUNT (ORANGES: INTEGER) return INTEGER;
function COUNT (APPLES: BIT) return INTEGER;
```

Both functions are overloaded since they have the same name, COUNT. When a call to either function is made, it is possible to identify the exact function to which the call is made from the type of actuals passed because they have different parameter types. For example, the function call

```
COUNT(20)
```

refers to the first function, since 20 is of type INTEGER, while the function call

```
COUNT('1')
```

refers to the second function, since the type of actual is BIT.

Here is another example.

```
function TO_CHARACTER (ARG: STD_LOGIC)
  return CHARACTER;
function TO_CHARACTER (ARG: STD_LOGIC_VECTOR)
  return CHARACTER;
```

Both functions are overloaded. The function call:

```
TO_CHARACTER ('0')
```

will call the first function since the argument is of type STD_LOGIC, while the function call:

```
TO_CHARACTER ("10010")
```

will call the second function since the argument is of type STD_LOGIC_VECTOR.

If two overloaded subprograms have the same parameter types and result types, it is possible for one subprogram to hide the other subprogram. This can happen if a subprogram is declared within another subprogram's scope. Here is an example.

```
architecture HIDING of DUMMY_ENTITY is
  function ADD (A, B: BIT_VECTOR) return BIT_VECTOR is
  begin
    -- Body of function here.
  end ADD;
begin
  SUM_IN_ARCH <= ADD(IN1, IN2);
  process
    function ADD (C, D: BIT_VECTOR) return BIT_VECTOR is
    begin
      -- Body of function here.
    end ADD;
  begin
    SUM_IN_PROCESS <= ADD (IN1, IN2);
    SUM_IN_ARCH <= HIDING.ADD(IN1, IN2);
  end process;
end HIDING;
```

The function ADD declared in the architecture body is hidden within the process because of the second function, ADD, declared within the declarative part of the process. This function can still be accessed by qualifying the function name with the architecture name, as shown in the second statement in the process.

It is also possible for two overloaded subprograms to be directly visible within a region, by using `use` clauses. In such a case, a subprogram call may be ambiguous, and hence an error, if it is not possible to determine which of the overloaded subprograms is being called. Here is an example.

```
package P1 is
  function ADD (A, B: BIT_VECTOR) return BIT_VECTOR;
end P1;

package P2 is
  function ADD (X,Y: BIT_VECTOR) return BIT_VECTOR;
end P2;

use WORK.P1.all, WORK.P2.all;
architecture OVERLOADED of DUMMY_ENTITY is
```

```
begin
   SUM_CORRECT <= ADD (X => IN1,Y => IN2);
   SUM_ERROR <= ADD (IN1, IN2);                    -- An error.
end OVERLOADED;
```

The function call in the first signal assignment statement is not an error, since it refers to the function declared in package P2 (the formal parameters are explicitly specified in the association), while the call in the second signal assignment statement is ambiguous and hence an error.

Two overloaded subprograms can also have a different number of parameters but have the same parameter types and result types. In this case, the number of actuals supplied in the subprogram call identifies the correct subprogram. Here is an example of such a set of functions that determine the smallest value from a set of 2, 4, or 8 integers.

```
function SMALLEST (A1, A2: INTEGER) return INTEGER;
function SMALLEST (A1, A2, A3, A4: INTEGER) return INTEGER;
function SMALLEST (A1, A2, A3, A4, A5, A6, A7, A8: INTEGER)
   return INTEGER;
```

A call such as:

```
... SMALLEST (4, 5) ...
```

refers to the first function, while the function call:

```
... SMALLEST (20, 45, 52, 1, 89, 67, 91, 22) ...
```

refers to the third function. This flexibility helps in writing code that is easy to decipher since the same subprogram name can be made to serve differently when used with a different set of inputs.

A call to an overloaded subprogram is ambiguous, and hence an error, if it is not possible to identify the exact subprogram being called using the following information:

- **i.** Subprogram name
- **ii.** Number of actuals
- **iii.** Types and order of actuals
- **iv.** Names of formal parameters (if named association is used)
- **v.** Result type (for functions)

Notice that overloading does not distinguish by subtypes. Here is an example to illustrate this point.

```
type LOGIC5 is ('X', '0', '1', 'D', 'Z');
subtype LOGIC5_01 is LOGIC5 range '0' to '1';

function RISING_EDGE (signal CLOCK: LOGIC5)
  return BOOLEAN;
function RISING_EDGE (signal CLOCK: LOGIC5_01)
  return BOOLEAN;
```

The second function declaration hides the first function, because both have the same parameter base types and the same return types. Thus, a call such as:

```
signal CK: LOGIC5_01;
. . .
. . . RISING_EDGE (CK) . . .
```

is an error.

Overloading also does not distinguish by the parameter modes.

```
procedure UNSIGNED_PLUS (A, B: in STD_LOGIC_VECTOR;
                           C: out STD_LOGIC_VECTOR);
procedure UNSIGNED_PLUS (A, B: in STD_LOGIC_VECTOR;
                           C: inout STD_LOGIC_VECTOR);
```

The second procedure hides the first procedure because both the procedures have the same parameter base types. The mode of the third parameter is different, but the mode is irrelevant when checking if two subprograms hide each other.

8.3 Operator Overloading

Operator overloading is one of the most useful features in the language. When a standard operator symbol is made to behave differently based on the type of its operands, the operator is said to be overloaded. The need for operator overloading arises from the fact that the predefined operators in the language are defined only for operands of certain predefined types. For example, the `and` operation is defined for arguments of type BIT and BOOLEAN, and for one-dimensional arrays of BIT and BOOLEAN only. What if the arguments were of type MVL (where MVL is a user-defined enumeration type with values 'U', '0', '1', and 'Z')? In such a case, it is possible to augment the and operation as a function that operates on arguments of type MVL. The **and** operator is then said to be overloaded. The operator in the expression:

```
S1 and S2
```

where S1 and S2 are of type MVL, would then refer to the and operation that was defined by the model writer as a function. The operator in the expression:

```
CLK1 and CLK2
```

where CLK1 and CLK2 are of type BIT, would refer to the predefined **and** operator.

Function bodies are written to define the behavior of overloaded operators. The number of parameters in such a function must match in cardinality with that of the predefined operator. The function thus has, at most, two parameters; the first one refers to the left, or the only, operand of the operator, and the second parameter, if present, refers to the second operand. Here are some examples of function declarations for such function bodies.

```
type MVL is ('U', '0', '1', 'Z');
function "and" (L, R: MVL) return MVL;
function "or" (L, R: MVL) return MVL;
function "not" (R: MVL) return MVL;
```

Since the **and**, **or**, and **not** operators are predefined operator symbols, they have to be enclosed within double quotes when used as overloaded operator function names. Having declared the overloaded functions, the operators can now be called using any of the following two different types of notations:

1. Standard operator notation
2. Standard function call notation

Here are some examples of these two types of notations based on the overloaded operator function declarations that appeared earlier.

```
signal A, B, C: MVL;
signal X, Y, Z: BIT;

A <= 'Z' or '1';                -- #1: standard operator notation.
B <= "or" ('0', 'Z');           -- #2: function call notation.
X <= not Y;                     -- #3
Z <= X and Y;                   -- #4
C <= (A or B) and (not C);      -- #5
Z <= (X and Y) or A;            -- #6
```

The **or** operator in the first statement refers to the overloaded operator because the type of the left operand is MVL. This is the standard operator notation since the overloaded operator symbol appears just like the standard operator symbol. An example of the function call notation is shown in the second statement in which the overloaded function, **or**, is explicitly called. The operators in the third and fourth statements refer to the predefined operators since their operands are of type BIT. The sixth statement would be an error, assuming that there are no overloaded **or** operators defined with the first parameter type of BIT and the second parameter type of MVL.

The last example brings up a very interesting point. In overloaded operator functions, it is not necessary for both operands to have the same type. In the previous case, if another **or** overloaded function with a declaration such as:

```
function "or" (L: BIT; R: MVL) return BIT;
```

were defined, the sixth assignment statement would not be an error.

It is often the case that a user deals mostly with vectors; therefore, it is necessary to provide a capability to perform arithmetic operations on vectors. Arithmetic operations are not predefined in the language to work on vectors. Overloaded functions again need to be defined. Here are some examples of overloaded operator declarations.

```
function "+" (OPD1, OPD2: BIT_VECTOR) return BIT_VECTOR;
function "–" (OPD1, OPD2: BIT_VECTOR) return BIT_VECTOR;

function "+" (OPD1, OPD2: STD_LOGIC_VECTOR)
    return STD_LOGIC_VECTOR;
function "–" (OPD1, OPD2: STD_LOGIC_VECTOR)
    return STD_LOGIC_VECTOR;
```

Examples of calls to these overloaded functions are given next.

```
variable A, B, C, CAB: BIT_VECTOR (3 downto 0);
variable D, E, F, FED: STD_LOGIC_VECTOR (0 to 7);
...
CAB := A + B – C;
FED := D – F + E;
```

In the first variable assignment statement, the arguments for the + and – operators are BIT vectors; therefore, these operations call the corresponding "+" and "–" overloaded functions for evaluation, respectively. Similarly, in the second variable assignment, the + and – operators operate on STD_LOGIC vectors; therefore, for evaluation, their corresponding overloaded operator functions for "+" and "–" are called.

8.4 Signatures

In earlier versions of the language, it was not possible to uniquely identify an overloaded subprogram or an overloaded enumeration literal. For example,

```
-- Two enumeration type declarations:
type MYBIT is ('0', '1');
type FOUR_VALUE is ('U', '0', '1', 'Z');

-- Two overloaded operator function declarations:
function "+" (A, B: BIT_VECTOR) return BIT_VECTOR;
function "+" (A, B: STD_LOGIC_VECTOR)
    return STD_LOGIC_VECTOR;
```

In the enumeration type declarations, '0' and '1' are overloaded enumeration literals, while in the two function declarations, "+" is an overloaded operator function. If there was a need to either define an alias for the first "+" or to assign an attribute (aliases and attributes are discussed in Chapter 10), it was not possible to uniquely distinguish either one of the "+" functions in earlier version of the language. Similarly, it was not possible to uniquely identify the '0' or '1' enumeration literal so that an attribute could be associated with or an alias created for either one of them.

Signatures have been provided as a means of distinguishing overloaded subprograms and overloaded enumeration literals. A *signature* shows the parameter types and result type of an overloaded subprogram or an overloaded enumeration literal. A signature is of the form:

```
[ first-parameter-type, second-parameter-type, . . .
    [ return function-return-type ] ]
```

Note that the outer square brackets ([ ]) are part of the syntax and do not represent optional items. Here are the signatures for the previously declared "+" overloaded functions.

```
[ BIT_VECTOR, BIT_VECTOR return BIT_VECTOR ]
[ STD_LOGIC_VECTOR, STD_LOGIC_VECTOR
    return STD_LOGIC_VECTOR ]
```

To determine the parameter and result types of an enumeration literal, an enumeration literal is treated as equivalent to a function with no parameters, with the function name being the same as the enumeration literal and the return type of the function being the enumeration type. For example, the equivalent functions for the '0' enumeration literals declared previously in type MYBIT and FOUR_VALUE are:

```
function '0' return MYBIT;
function '0' return FOUR_VALUE;
```

Therefore, the signatures for each of these enumeration literals are:

```
[ return MYBIT ]
[ return FOUR_VALUE ]
```

Signatures can be used in alias declarations, attribute specifications, and attribute names to uniquely identify an overloaded subprogram or an overloaded enumeration literal. Examples of these are given in Chapter 10.

8.5 Default Values for Parameters

It is possible to assign default values to input parameters in a subprogram declaration. In such a case, if no actual is specified for a formal parameter that has a default value, then the default value is used as a value for the formal.

Here is an example of such a specification.

```
function MY_AND (A: in BIT := '0', B: in BIT := '1') return BIT;
```

When a default value is specified for a parameter, then it is optional to pass an actual value in a subprogram call. When no actual is passed in for such a parameter (or the keyword **open** is used as an actual), the parameter assumes the default value. The default value is ignored if an actual value (other than keyword **open**) is explicitly specified. Here are some examples of function calls.

```
. . . MY_AND ( ) . . .
  -- Formal parameters A and B have the default values
  -- of '0' and '1' respectively.

. . . MY_AND (B => SIG_P) . . .
  -- SIG_P is passed in as the value for B. Since no actual is
  -- specified for A, A has its default value '0'.

. . . MY_AND (B => SIG_P, A => SIG_Q) . . .
  -- Both the default values are ignored since actuals are
  -- explicitly specified.

. . . MY_AND (A => SIG_Q, B => open) . . .
  -- Using keyword open as an actual is equivalent to not specifying
  -- a value; thus, B has a default value of '1'.
```

Here is another example. This function can convert an integer value to a value of type TIME.

```
function TO_TIME (ARG: INTEGER; SCALE: TIME := ns)
    return TIME is
begin
    return ARG * SCALE;
end TO_TIME;
```

Notice that a default value has been specified for the formal parameter SCALE of 1 ns. A function call such as:

```
TO_TIME (56)
```

uses a SCALE of 1 ns, while a function call:

```
TO_TIME (56, 10 ms)
```

uses a SCALE of 10 ms; the default value is ignored in this case because an actual value has been explicitly specified.

An input signal parameter in a subprogram cannot have a default value associated with it. For example,

```
procedure TO_VECTOR (signal VALUE: in INTEGER; . . .);
```

VALUE is a signal parameter and cannot have a default value associated with it.

❑

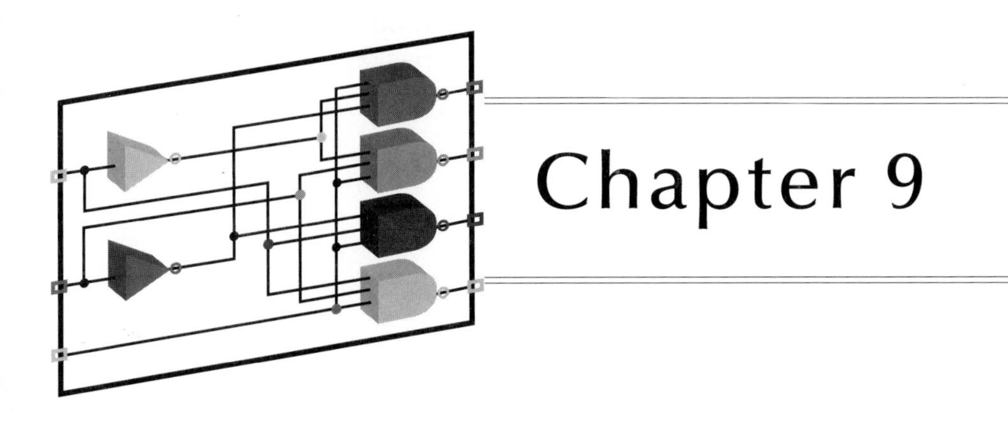

Chapter 9

Packages and Libraries

This chapter explains packages and how compiled design units are stored in design libraries. It explains how the contents of design units stored in different libraries may be shared by several design units.

A package provides a convenient mechanism to store and share declarations that are common across many design units. A package is represented by:

1. a package declaration and, optionally,
2. a package body.

9.1 Package Declaration

A package declaration contains a set of declarations that may possibly be shared by many design units. It defines the interface to the package, that is, it defines items that can be made visible to other design units, for example, a function declaration. A package body, in contrast, contains the hidden details of a package, for example, a function body.

The syntax of a package declaration is:

```
package package-name is
  package-item-declarations --> These may be:
    -- subprogram declarations
    -- type declarations
    -- subtype declarations
    -- constant declarations
    -- signal declarations
    -- variable declarations
    -- file declarations
    -- alias declarations
    -- component declarations
    -- attribute declarations
    -- attribute specifications
    -- disconnection specifications
    -- use clauses
end [ package ] [ package-name ] ;
```

An example of a package declaration is given next.

```
package SYNTH_PACK is
  constant LOW2HIGH: TIME := 20 ns;
  type ALU_OP is (ADD, SUB, MUL, DIV, EQL);
  attribute PIPELINE: BOOLEAN;
  type MVL is ('U', '0', '1', 'Z');
  type MVL_VECTOR is array (NATURAL range <>) of MVL;
  subtype MY_ALU_OP is ALU_OP range ADD to DIV;
  component NAND2
    port (A, B: in MVL; C: out MVL);
  end component;
end SYNTH_PACK;
```

Items declared in a package declaration can be accessed by other design units by using the `library` and `use` clauses. The set of common declarations may also include function and procedure declarations and deferred constant declarations. In this case, the behavior of the subprograms and the values of the deferred constants are specified in a separate design unit called the package body. Since the previous package example does not contain any subprogram declarations or deferred constant declarations, a package body is not necessary.

Consider the following package declaration.

```
use WORK.SYNTH_PACK.all;            -- Includes all declarations
                                    -- from package SYNTH_PACK.
package PROGRAM_PACK is
  constant PROP_DELAY: TIME;          -- A deferred constant.
  function "and" (L, R: MVL) return MVL;
  procedure LOAD (signal ARRAY_NAME: inout MVL_VECTOR;
    START_BIT, STOP_BIT, INT_VALUE: in INTEGER);
end package PROGRAM_PACK;           -- The keyword package after end
                                    -- is optional.
```

In this case, a package body is required because the package declaration contains a deferred constant declaration and two subprogram declarations.

See Appendix E for a listing of the package declaration for the IEEE standard package STD_LOGIC_1164.

9.2 Package Body

A package body primarily contains the behavior of the subprograms and the values of the deferred constants declared in a package declaration. It also may contain other declarations, as shown by the following syntax of a package body.

```
package body package-name is
  package-body-item-declarations --> These are:
    -- subprogram bodies
    -- complete constant declarations
    -- subprogram declarations
    -- type and subtype declarations
    -- file and alias declarations
    -- use clauses
end [ package body ] [ package-name ] ;
```

The package name must be the same as the name of its corresponding package declaration. A package body is not necessary if its associated package declaration does not have any subprogram or deferred constant declarations. The associated package body for the package declaration PROGRAM_PACK, described in the previous section, is:

```
package body PROGRAM_PACK is
  use WORK.TABLES.all;
```

```
    constant PROP_DELAY: TIME := 15 ns;
    function "and" (L, R: MVL) return MVL is
    begin
      return TABLE_AND(L, R);
      -- TABLE_AND is a 2-D constant defined in another package,
      -- TABLES.
    end "and";

    procedure LOAD (signal ARRAY_NAME: inout MVL_VECTOR;
        START_BIT, STOP_BIT, INT_VALUE: in INTEGER) is
      -- Local declarations here.
    begin
      -- Procedure behavior here.
    end LOAD;
  end PROGRAM_PACK;
```

An item declared inside a package body has its scope restricted to be within the package body, and this item cannot be made visible in other design units. This restriction is in contrast to items declared in a package declaration, which can be accessed by other design units. Therefore, a package body is used to store private declarations that should not be visible, while a package declaration is used to store public declarations, which other design units can access. This is very similar to declarations within an architecture body, which are not visible outside of its scope, while items declared in an entity declaration can be made visible to other design units. An important difference between a package declaration and an entity declaration is that an entity can have multiple architecture bodies with different names, while a package declaration can have at most one package body, the names for both being the same.

9.3 Design File

A *design file* is an ASCII file containing VHDL source. It can contain one or more design units, where a design unit is one of the following:

- Entity declaration
- Architecture body
- Configuration declaration
- Package declaration
- Package body

This means that each design unit can also be placed in a separate file.

Figure 9-1 shows the compilation process. The design file is processed by a VHDL analyzer, which, after verifying the syntactic and semantic correctness of the source, compiles each design unit present in the design file into an intermediate form. Each intermediate form is stored in a design library that has been designated as the working library.

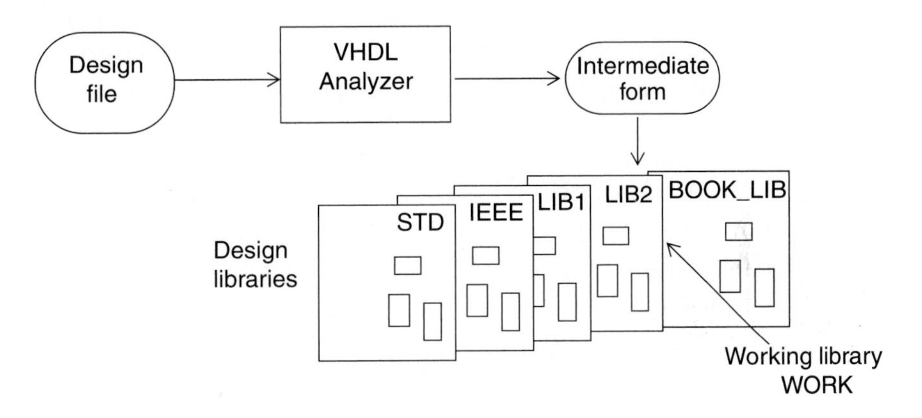

Figure 9-1 A typical compilation process.

9.4 Design Libraries

A compiled design unit is stored in a design library. A design library is an area of storage in the file system of the host environment. The format of this storage is not defined by the language. In one scheme, a design library is implemented on a host system as a file directory, and the compiled design units are stored as files in this directory. The management of the design libraries is also not defined by the language and is again tool-implementation-specific.

An arbitrary number of design libraries may be specified. Each design library has a logical name with which it is referenced inside a VHDL description. The association of the logical names with their physical storage names is maintained by the host environment. There is one design library with the logical name STD predefined by the language; this library contains the compiled descriptions for the two predefined packages STANDARD and TEXTIO. Exactly one design library must be designated as the working library with the logical name WORK. When a design file is compiled, the intermediate forms of the design units present in the design file are always stored in the working library. Therefore, before compilation begins, the logical name WORK must point to one of the design libraries (see Figure 9-1).

There may exist, optionally, another design library called IEEE. This design library contains the package STD_LOGIC_1164, which defines a nine-value logic type and its associated overloaded functions and other utilities. This package is an IEEE standard, IEEE Std 1164-1993, developed for VHDL model interoperability. Note that this package and the IEEE design library are not part of the VHDL language standard.

The VHDL source is present in a design file; a design file can contain one or more design units. Design units are further classified as following:

1. *Primary* units: These units are not directly dependent on other design units. They are:
   - **a.** *Entity declarations*: The items declared in an entity declaration are implicitly visible within the associated architecture bodies.
   - **b.** *Package declarations*: Items declared within a package declaration can be exported to other design units using `library` and `use` clauses. Items declared within a package declaration are also implicitly visible within the corresponding package body.
   - **c.** *Configuration declarations*.
2. *Secondary* units: These units are always dependents of a primary unit. These design units do not allow items declared within them to be exported out of the design unit, that is, the items cannot be referenced in other design units. These secondary units are:
   - **a.** *Architecture bodies*: For example, a signal declared in an architecture body cannot be referenced in other design units.
   - **b.** *Package bodies*.

There can be exactly one primary unit with a given name in a single design library. Secondary units associated with different primary units can have identical names in the same design library; also, a secondary unit may have the same name as its associated primary unit. For example, assume there exists an entity called AND_GATE in a design library. It may have an architecture body with the same name, and another entity MY_GATE in the same design library may have an architecture body that also has the name AND_GATE.

Secondary units must coexist with their associated primary units in the same design library; for example, an entity declaration and all of its architecture bodies must reside in the same library. Similarly, a package declaration and its associated package body must reside in a single library.

Even though a configuration declaration is a primary unit, it must reside in the same library as the entity declaration that it configures.

9.5 Order of Analysis

Since it is possible to export items declared in primary units to other design units, a constraint is imposed in the sequence in which design units must be analyzed. A design unit that references an item declared in another primary unit can be analyzed only after that primary unit has been analyzed. For example, if a configuration declaration references an entity COUNTER, the entity declaration for COUNTER must be analyzed before the configuration declaration.

A primary unit must be analyzed before any of its associated secondary units. For example, an entity declaration must be analyzed before its architecture bodies are analyzed.

9.6 Implicit Visibility

An architecture body implicitly inherits all declarations in the entity since it is tied to that entity by virtue of the statement:

architecture *architecture-name* **of** *entity-name* **is** . . .

Similarly, a package body implicitly inherits all items declared in the package declaration by virtue of its first statement:

package body *package-name* **is** . . .

where the package name is the same as the one in the package declaration.

9.7 Explicit Visibility

Explicit visibility of items declared in other design units can be achieved using the following two clauses:

1. `library` clause
2. `use` clause

A `use` clause can appear in any declarative part of a design unit. If a `library` clause or a `use` clause appears at the beginning of a design unit, it is called a *context clause*. Figure 9-2 shows an example. Items specified in the context clause become visible only to the design unit that follows the context clause; the items are not visible to other primary design units that may be present in the same design file. This means that if a design file contains three primary design units, such as in the example shown in Figure 9-3, context clauses must be specified for each design unit, if needed. For example, the context clauses specified before design unit A are visible to design unit A only; context clauses specified before design unit B are visible to design unit B only.

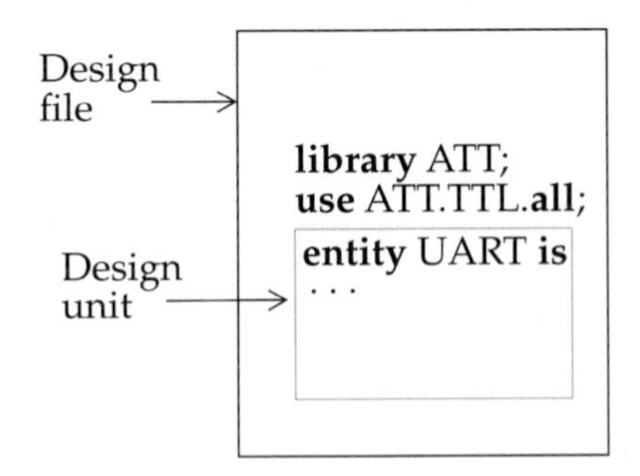

Figure 9-2 Context clause associated with the following design unit.

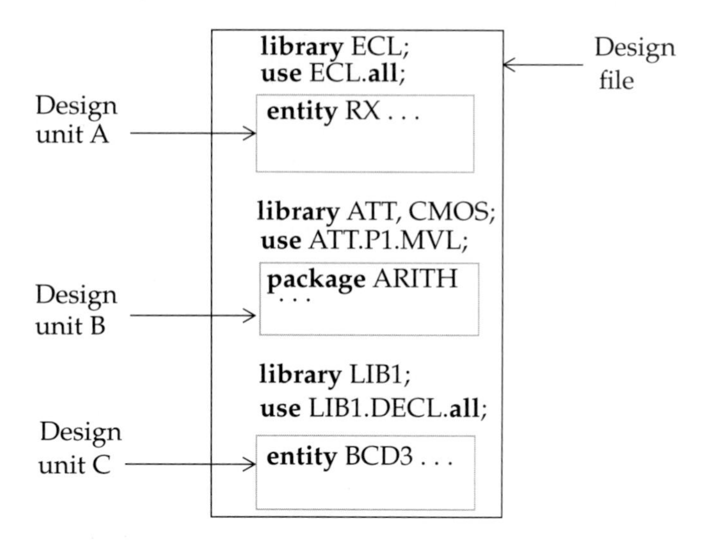

Figure 9-3 Separate context clauses with each design unit.

9.7.1 Library Clause

The `library` clause makes visible the logical names of design libraries that can be referenced within a design unit. The format of a `library` clause is:

library *list-of-logical-library-names* ;

The `library` clause:

library TTL, CMOS;

makes the logical names TTL and CMOS visible in the design unit that follows. Note that the `library` clause does not make design units or items present in the library visible, it makes only the library name visible (it is like a declaration for a library name). For example, it would be illegal to use the name "TTL.SYNTH_PACK.MVL" within a design unit without first declaring the library name using the "**library** TTL;" clause.

The `library` clause:

library STD, WORK;

is implicitly declared for every design unit.

9.7.2 Use Clause

There are two main forms of the use clause.

```
use library-name . primary-unit-name ;              -- Form 1.
use library-name . primary-unit-name . item ;       -- Form 2.
```

The first form of the use clause allows the specified primary unit name from the specified design library to be referenced in a design description. For example,

```
library CMOS;
use CMOS.NOR2;
configuration . . . is
      . . . use entity NOR2( . . . );
end;
```

Note that entity NOR2 must be available in compiled form in the design library CMOS before attempting to compile the design unit where it is used.

The second form of the use clause makes the item declared in the primary unit visible; the item can therefore be referenced within the following design unit. For example,

```
library IEEE;
use IEEE.STD_LOGIC_1164.STD_LOGIC;
-- STD_LOGIC is a type declared in STD_LOGIC_1164 package.
-- The package STD_LOGIC_1164 is stored in the design library IEEE.
entity NAND2 is
   port (A, B: in STD_LOGIC; . . . ) . . .
```

If all items within a primary unit are to be made visible, the keyword **all** can be used. For example,

```
use IEEE.STD_LOGIC_1164.all;
```

makes all items declared in package STD_LOGIC_1164 in design library IEEE visible. The use clause:

```
use IEEE.all;
```

makes all names of primary units present in the design library IEEE visible.

Items external to a design unit can be accessed by other means as well. One way is to use a selected name. An example of using a selected name is:

```
library IEEE;
use IEEE.STD_LOGIC_1164;
entity NOR2 is
  port (A, B: in STD_LOGIC_1164.STD_LOGIC; . . . ) . . .
```

Since only the primary unit name was made visible by the `use` clause, the primary unit name must be used along with the name of the type being referenced, that is, STD_LOGIC_1164.STD_LOGIC must be specified. Another example is shown next. The type VALUE_9 is defined in package SIMPACK, which has been compiled into the CMOS design library.

```
library CMOS;
package P1 is
  procedure LOAD (A, B: CMOS.SIMPACK.VALUE_9; . . . ) ;
  . . .
end P1;
```

In this case, the primary unit name SIMPACK was specified only at the time of usage.

So far, we talked about exporting items across design libraries. What if it is necessary to export items from design units that are in the same library? In this case, there is no need to specify a `library` clause since every design unit has the following `library` clause implicitly declared.

```
library STD, WORK;
```

The predefined design library STD contains the packages STANDARD and TEXTIO. The package STANDARD contains declarations for predefined types such as CHARACTER, BOOLEAN, BIT_VECTOR, and INTEGER. The following `use` clause is also implicitly declared for every design unit.

```
use STD.STANDARD.all;
```

Thus, all items declared within the package STANDARD are available for use in every VHDL description. However, there is no such implicit declaration for the TEXTIO package. If items declared in this package need to be referenced, a `use` clause of the following form may be used.

```
use STD.TEXTIO.all;
```

❑

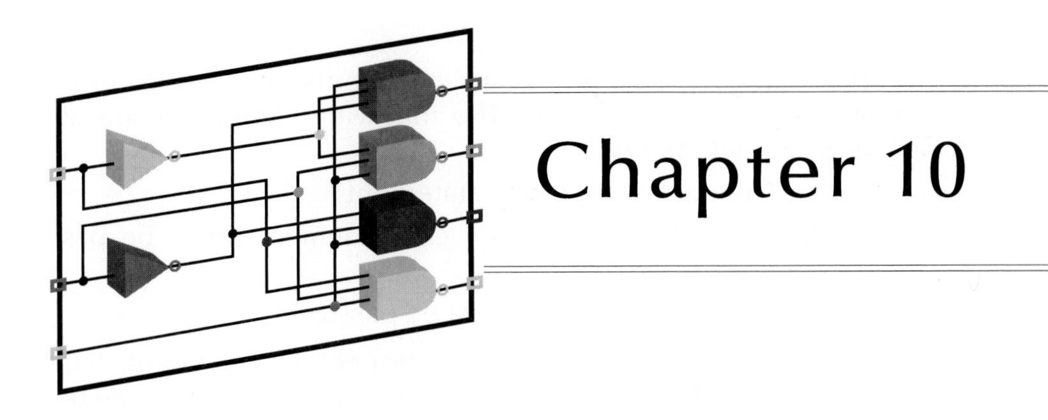

Chapter 10

Advanced Features

This chapter describes some of the more general features of the language that cannot be characterized as belonging to any specific modeling style. These include, among others, attributes, type conversions, entity statements, aliases, generate statements, and guarded signals. The usage of block statements as a partitioning mechanism is also described.

10.1 Entity Statements

Certain statements that are common to all architecture bodies of an entity can be inserted into the entity declaration. Common declarations appear in the entity declarative part, while other common statements appear in the entity statement part, as shown.

```
entity entity-name is
   [ generic ( . . . ); ]
   [ port ( . . . ); ]
   [ entity-item-declarations ]                    -- Declarative part.
[ begin
```

```
    entity-statements ]                                -- Statements part.
  end [ entity ] [ entity-name ] ;
```

Entity-statements must be *passive* statements, that is, they must not assign values to any signals. The following statements are allowed as entity statements.

1. Concurrent assertion statement
2. Concurrent procedure call statement (must be passive)
3. Process statement (must be passive)

Entity statements can be used to monitor certain operating characteristics of an entity. For example, in an RS flip-flop, a check can be made to ensure that signals R and S are never high simultaneously. Here is a way of modeling this using an assertion statement.

```
entity RS_FLIPFLOP is
  port (R,S: in BIT; Q, QBAR: out BIT);
  constant FF_DELAY: TIME := 24 ns;
  type FF_STATE is (ONE, ZERO, UNKNOWN);
begin
  assert not (R = '1' and S = '1')
    report "Not valid inputs!"
    severity ERROR;
end entity RS_FLIPFLOP;  -- The keyword entity after end is optional.
```

The constant and type declarations in this example are examples of item declarations within an entity declaration. Items declared by these declarations are visible to all architecture bodies associated with this entity declaration.

In the following example, the timing of a D flip-flop is checked using a concurrent procedure call that appears as an entity statement. This procedure will be called every time there is an event on either of the input ports D or CLK, irrespective of the contents in the architecture bodies that are associated with this entity.

```
use WORK.MYPACK.CHECK_SETUP;
entity DFF is
  port (D, CLK: in BIT; Q, QBAR: out BIT);
  constant SETUP: TIME := 7 ns;
begin
  CHECK_SETUP (D, CLK, SETUP);
end DFF;
```

10.2 Generate Statements

Concurrent statements can be conditionally selected or replicated during the elaboration phase using the generate statement. There are two forms of the generate statement.

1. Using the for-generation scheme, concurrent statements can be replicated a predetermined number of times.
2. With the if-generation scheme, concurrent statements can be conditionally elaborated.

The generate statement is interpreted during elaboration, and therefore has no simulation semantics. It resembles a macro expansion. The generate statement provides for a compact description of regular structures such as memories, registers, and counters.

10.2.1 For-generation Scheme

The format of a generate statement using the for-generation scheme is:

generate-label : **for** *generate-identifier* **in** *discrete-range* **generate**
  [*block-declarations*
begin]
  concurrent-statements
end generate [*generate-label*] ;

The values in the discrete range must be globally static, that is, they must be computable at elaboration time. During elaboration, the set of concurrent statements are replicated once for each value in the discrete range. These statements can also use the generate identifier in their expressions, and its value would be substituted during elaboration for each replication. There is an implicit declaration for the generate identifier within the generate statement; therefore, no declaration for this identifier is required. The type of the identifier is defined by the discrete range. Declarations, if present, declare items that are visible only within the generate statement.

Consider the following representation of a 4-bit full-adder, shown in Figure 10-1, using the generate statement.

```
entity FULL_ADD4 is
  port (A, B: in BIT_VECTOR(3 downto 0); CIN: in BIT;
    SUM: out BIT_VECTOR(3 downto 0); COUT: out BIT);
end FULL_ADD4;

architecture FOR_GENERATE of FULL_ADD4 is
  component FULL_ADDER
    port (PA, PB, PC: in BIT; PCOUT, PSUM: out BIT);
  end component;
  signal CAR: BIT_VECTOR(4 downto 0);
begin
  CAR(0) <= CIN;
  GK: for K in 3 downto 0 generate
    FA: FULL_ADDER port map (CAR(K), A(K), B(K), CAR(K+1), SUM(K));
  end generate GK;
  COUT <= CAR(4);
end FOR_GENERATE;
```

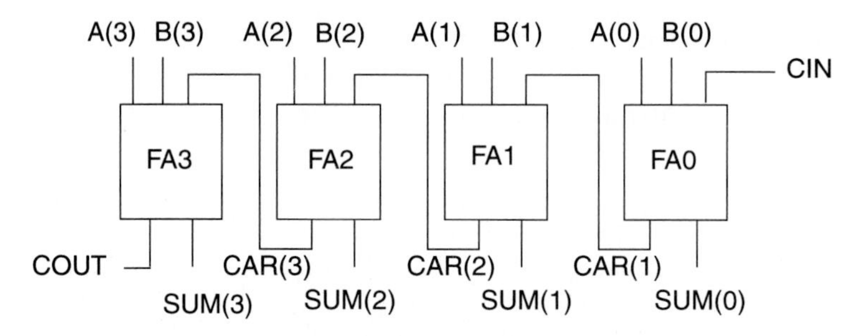

Figure 10-1 A 4-bit full-adder.

During elaboration, the generate statement is expanded to the following four block statements.

```
GK: block
  constant K: INTEGER := 3;
begin
  FA: FULL_ADDER
      port map (CAR(K), A(K), B(K), CAR(K+1), SUM(K));
end block GK;

GK: block
  constant K: INTEGER := 2;
begin
  FA: FULL_ADDER
```

```
        port map (CAR(K), A(K), B(K), CAR(K+1), SUM(K));
end block GK;

GK: block
   constant K: INTEGER := 1;
begin
   FA: FULL_ADDER
        port map (CAR(K), A(K), B(K), CAR(K+1), SUM(K));
end block GK;

GK: block
   constant K: INTEGER := 0;
begin
   FA: FULL_ADDER
        port map (CAR(K), A(K), B(K), CAR(K+1), SUM(K));
end block GK;
```

The body of the generate statement can also have other concurrent statements. For example, in the previous architecture body the component instantiation statement could be replaced by concurrent signal assignment statements, as shown next.

```
G2: for M in 3 downto 0 generate
   SUM(M) <= A(M) xor B(M) xor CAR(M);
   CAR(M+1) <= A(M) and B(M) and CAR(M);
end generate G2;
```

10.2.2 If-generation Scheme

The second form of the generate statement uses the if-generation scheme. The format for this type of generate statement is:

```
generate-label : if expression generate
   [ block-declarations
begin ]
   concurrent-statements
end generate [ generate-label ] ;
```

The if-generate statement allows for conditional selection of concurrent statements based on the value of an expression. This expression must be a globally static expression, that is, the value must be computable at elaboration time. Any declarations present are again local to the generate statement.

Here is an example of a 4-bit counter, shown in Figure 10-2, modeled using the if-generate statement.

```
entity COUNTER4 is
  port (COUNT, CLOCK: in BIT; Q: buffer BIT_VECTOR(0 to 3));
end COUNTER4;

architecture IF_GENERATE of COUNTER4 is
  component D_FLIP_FLOP
    port (D, CLK: in BIT; QUE: out BIT);
  end component;
begin
  GK: for K in 0 to 3 generate
    GK0: if K = 0 generate
      DFF: D_FLIP_FLOP port map (COUNT, CLOCK, Q(K));
    end generate GK0;
    GK1_3: if K > 0 generate
      DFF: D_FLIP_FLOP port map (Q(K–1), CLOCK, Q(K));
    end generate GK1_3;
  end generate GK;
end IF_GENERATE;
```

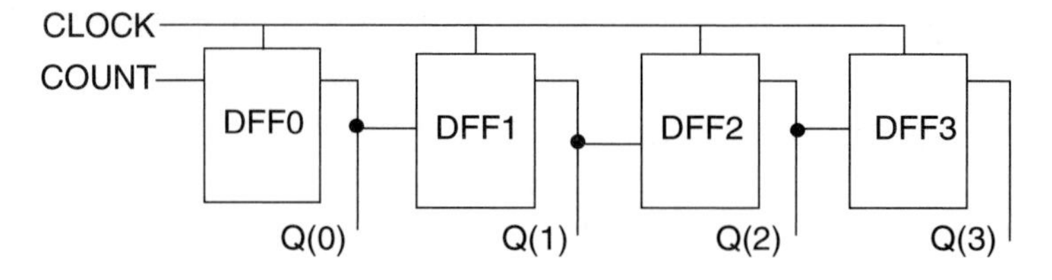

Figure 10-2 A 4-bit counter.

Here are the generate statements for an example that selects a buffer with different delays based on the value of a constant.

```
GA: if USER_WANTS = LOW_DELAY generate
  Z <= A after 2 ns;
end generate;
GB: if USER_WANTS = MEDIUM_DELAY generate
  Z <= A after 10 ns;
end generate;
GC: if USER_WANTS = HIGH_DELAY generate
  Z <= A after 25 ns;
end generate;
```

The if-generate statement is useful in modeling repetitive structures, especially in modeling boundary conditions. An if-generate statement does not have an `else` or an `elsif` branch.

10.2.3 Binding Component Instances

Component instances in a generate statement can be bound to entities using a block configuration. A block configuration is defined for each range of generate labels. Here is an example of such a binding using a configuration declaration (for entity FULL_ADD4 defined in Section 10.2.1).

```
configuration GENERATE_BIND of FULL_ADD4 is
  use WORK.all;                  -- Example of a declaration in the
                                 -- configuration declarative part.
  for FOR_GENERATE               -- A block configuration.
    for GK(1)                    -- A block configuration.
      for FA: FULL_ADDER
        use configuration WORK.FA_HA_CON;
      end for;
    end for;
    for GK(2 to 3)
      for FA: FULL_ADDER           -- No explicit binding.
        -- Use defaults, that is, use entity FULL_ADDER
        -- in working library.
      end for;
    end for;
    for GK(0)
      for FA: FULL_ADDER
        use entity WORK.FULL_ADDER (FA_DATAFLOW);
      end for;
    end for;
  end for;
end GENERATE_BIND;
```

There are three block configurations that apply to the generate statement, one each for GK(1), GK(2 **to** 3), and GK(0). Each of these block configurations define the bindings for the components valid for that generate index.

Components can also be bound using configuration specifications. This is done by specifying the configuration specifications in the declarative part of the generate statement. Note that the specifications cannot appear in the enclosing architecture declarative part since a generate statement forms a separate declarative region. Here is the generate statement that appears in entity FULL_ADD4 with the configuration specification.

```
GK: for K in 3 downto 0 generate
  -- Configuration specification:
```

```
    for all: FULL_ADDER use entity WORK.FA_CMOS;
    -- Bind all instances of component FULL_ADDER appearing
    -- within the generate statement to entity FA_CMOS residing in
    -- library WORK.
begin           -- Note the use of keyword begin when declarations
                -- are present in a generate statement.
    FA: FULL_ADDER port map  (CAR(K), A(K), B(K),
                              CAR(K+1), SUM(K));
end generate GK;
```

In the IEEE Std 1076-1987 version of the language, a declarative part was not allowed within a generate statement. In such a case, if components need to be bound to instances using configuration specifications, a dummy block statement can be used. This is shown using an example.

```
GK: for K in 3 downto 0 generate
    LBL_A: block
        -- Configuration specification:
        for all: FULL_ADDER use entity WORK.FA_CMOS;
        -- Bind all instances of component FULL_ADDER appearing
        -- within the generate statement to entity FA_CMOS residing in
        -- library WORK.
    begin
        FA: FULL_ADDER port map  (CAR(K), A(K), B(K),
                                  CAR(K+1), SUM(K));
    end block LBL_A;
end generate GK;
```

The block statement is used here to provide a declarative region so that configuration specifications can be specified.

10.3 Aliases

An alias declares an alternate name for all or part of a named item. It provides a convenient shorthand notation for items that have long names. It also provides a mechanism to refer to the same named item in different ways depending on the context. For example,

```
signal S: BIT_VECTOR (31 downto 0);
```

can represent:

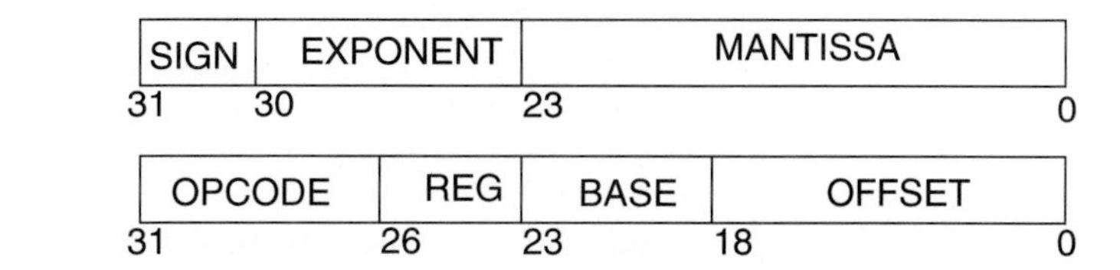

or:

Signal S can be thought of as being composed of a 24-bit mantissa, a 7-bit exponent, and a 1-bit sign. The same signal S can be thought of as being composed of a 19-bit offset, a 5-bit base, a 3-bit register code, and a 5-bit opcode. An alias can be defined for each of the required fields.

The syntax for an alias declaration is:

```
alias identifier [ : identifier-type ] is item-name ;
```

A named item can be an object, that is, a constant, signal, variable, or file. An alias of an object is called an *object alias*. An alias of other named items, such as function names, literals, type names, and attribute names, is called a *non-object alias*. An alias can be declared for any named item except labels, loop parameters, and generate parameters.

Here are some examples of object aliases.

```
variable DATA_WORD: BIT_VECTOR(15 downto 0);
alias DATA_BUS: BIT_VECTOR(7 downto 0) is
        DATA_WORD(15 downto 8);
alias STATUS: BIT_VECTOR(0 to 3) is DATA_WORD(3 downto 0);
alias RESET: BIT is DATA_WORD(4);
alias RX_READY: BIT is DATA_WORD(5);
```

Given these declarations, DATA_BUS can be used wherever DATA_WORD(15 **downto** 8) is needed, RESET can be used wherever DATA_WORD(4) is needed, and so on. It is important to note that assigning a value to an alias name is the same as assigning a value to the aliased name. The alias does not declare a new object but merely provides an alternate way of referring to the original object. The alias of an array object may change the way in which the array is indexed, as shown in the alias declaration of STATUS.

No type information needs to be specified for a non-object alias. However, a signature is required for an alias of a subprogram or an enumeration literal. Here are some examples of non-object aliases.

```
function TO_STDLOGICVECTOR (A: INTEGER; SIZE: INTEGER)
   return STD_LOGIC_VECTOR;
function "mod" (A, B: STD_LOGIC_VECTOR)
   return STD_LOGIC_VECTOR;

alias MOD_SLV is
   "mod" [ STD_LOGIC_VECTOR, STD_LOGIC_VECTOR
   return STD_LOGIC_VECTOR ];
alias TO_S is TO_STDLOGICVECTOR [ INTEGER, INTEGER
   return STD_LOGIC_VECTOR ];

alias ZERO is FALSE [ return BOOLEAN ];

type STD_LOGIC_VECTOR is array (NATURAL range <>)
   of STD_LOGIC;
alias SIGNED is STD_LOGIC_VECTOR;
```

The first two alias declarations declare aliases for functions TO_STDLOGICVECTOR and "**mod**". Signatures for these functions are explicitly specified in their names. The third alias declaration declares an alias ZERO for the BOOLEAN enumeration literal FALSE. Notice that in this case, we again need to explicitly specify the signature of the enumeration literal. The fourth alias declaration aliases a type; any use of SIGNED is the same as using STD_LOGIC_VECTOR.

Overloaded subprograms and overloaded enumeration literals are distinguished by their signatures. The following examples show two type declarations that have '0' as an overloaded enumeration literal.

```
type FOUR_VALUE is ('U', '0', '1', 'Z');
type STD_ULOGIC is ('U', 'X', '0', '1', 'Z', 'W', 'L', 'H', '-');
```

An alias can be defined for the '0' overloaded enumeration literals as follows.

```
alias STD0 is '0' [ return STD_ULOGIC ];
alias FV0 is '0' [ return FOUR_VALUE ];
```

STD0 is an alias for the enumeration literal '0' declared in type STD_ULOGIC. FV0 is the alias for literal '0' declared in type FOUR_VALUE. Note that the signature of the enumeration literal is specified to distinguish between the two overloaded enumeration literals.

Here are some examples of aliases of overloaded procedures.

```
procedure PRESET_CLEAR (signal DRV: STD_LOGIC_VECTOR;
   PC_VALUE: INTEGER);
```

```
procedure PRESET_CLEAR (signal DRV: BIT_VECTOR;
   PC_VALUE: INTEGER);

alias PCM is PRESET_CLEAR [ STD_LOGIC_VECTOR, INTEGER ];
alias PCB is PRESET_CLEAR [ BIT_VECTOR, INTEGER ];
```

PCM is the alias for the first procedure, while PCB is the alias for the second procedure. Signatures are again used to distinguish between the two.

10.4 Qualified Expressions

The type of an expression can be explicitly specified by qualifying the expression with its type. For example,

```
INTEGER'(A * B + C)
```

qualifies the expression (A * B + C) to have an integer value. The qualification is useful for type checking, and it does not imply any type conversion.

Apart from its type-checking usefulness, type qualification can be used to qualify types for expressions whose types cannot be determined from their context. Consider the following two overloaded procedure declarations.

```
procedure CHAR2INT (A: in CHARACTER; Z: out INTEGER);
procedure CHAR2INT (A: in BIT; Z: out INTEGER);
```

A call to a procedure such as:

```
CHAR2INT ('1', N);
```

is ambiguous, since it is not clear which overloaded procedure is to be called. However, if the expression in the procedure call were modified to be:

```
CHAR2INT (CHARACTER'('1'), N);
```

it is clear that this call refers to the first procedure.

Here is another example.

```
function "*" (A: STD_LOGIC_VECTOR; B: STD_LOGIC_VECTOR)
   return STD_LOGIC_VECTOR;
function "*" (A: STD_LOGIC_VECTOR; B: BIT_VECTOR)
   return STD_LOGIC_VECTOR;
```

A function call of the form:

```
RY * "00001"                    -- RY is of type STD_LOGIC_VECTOR
```

is ambiguous, since the type of the second operand could be either STD_LOGIC_VECTOR or BIT_VECTOR. To make the function call unambiguous, the second operand can be qualified to make the intent clear, such as:

```
RY * STD_LOGIC_VECTOR'("00001")
```

which now refers to the first "*" overloaded operator function.

Consider another example.

```
function TO_INTEGER (OPD: BIT_VECTOR)
  return INTEGER;
function TO_INTEGER (OPD: STD_LOGIC_VECTOR)
  return INTEGER;
```

A function call of the form:

```
TO_INTEGER ("10001")
```

is ambiguous since the type of the value being passed could either be a BIT_VECTOR or a STD_LOGIC_VECTOR. To make this function call unambiguous, the expression needs to be qualified by the type, such as:

```
TO_INTEGER (BIT_VECTOR'("10001"))
```

which now explicitly refers to the first TO_INTEGER function.

Here is a commonly occurring problem that requires a qualified expression. In the predefined package TEXTIO, the WRITE procedure is overloaded on both BIT_VECTOR and STRING.

```
procedure WRITE (L: inout LINE; VALUE: in BIT_VECTOR;
                JUSTIFIED: in SIDE := RIGHT; FIELD: in WIDTH := 0);
procedure WRITE (L: inout LINE; VALUE: in STRING;
                JUSTIFIED: in SIDE := RIGHT; FIELD : in WIDTH := 0);
```

A procedure call such as:

```
WRITE (L, "1110001");
```

is ambiguous and hence an error, since it is not possible to figure out if the argument "1110001" is a BIT_VECTOR or a STRING. The argument needs to be type qualified such as:

```
WRITE (L, BIT_VECTOR'("1110001"));
```

It is clear that in this procedure call, the WRITE procedure refers to the first one, the one with the BIT_VECTOR argument.

10.5 Type Conversions

The language does allow for very restricted type casting, that is, explicitly converting values between types. For example,

```
SUM := INTEGER (POLYWIDTH * 1.5);
```

will convert the real value obtained from POLYWIDTH * 1.5 to its equivalent integer value and assign it to SUM.

Type conversions are allowed between closely related types. These include those between integer and real, between array types that have the same dimensions and whose index types are closely related and element types are the same. Any other kind of type conversion, other than those predefined by the language, can be performed by using a user-defined function.

Here are some more examples of type conversions.

```
type SIGNED is array (NATURAL range <>) of BIT;
type BIT_VECTOR is array (NATURAL range <>) of BIT;
signal FCR: SIGNED (0 to 7);
signal EMA: BIT_VECTOR (0 to 7);
```

Without type conversions,

```
FCR <= EMA;
```

is illegal, since signals FCR and EMA are of different types. However, using type conversion,

```
FCR <= SIGNED (EMA);
```

makes the assignment legal. Type conversion is allowed in this case since types SIGNED and BIT_VECTOR are closely related, that is, they are both vectors of the same element type, and their index types are closely related.

To get a real value out of an integer division, apply type conversion to real on both the numerator and denominator before the division. For example,

```
signal REAL_SIG: REAL;
signal INT_A, INT_B: INTEGER;
. . .
REAL_SIG <= REAL (INT_A) / REAL(INT_B);
```

The integers INT_A and INT_B are converted to real values before division is performed which produces a real value result.

10.6 Guarded Signals

A *guarded* signal is a special type of signal that is declared to be of a **register** or **bus** kind in its declaration. A general form of a signal declaration is:

```
signal list-of-signals : [ resolution-function ] signal-type
                          [ signal-kind ] [ := expression ] ;
```

A guarded signal must be a resolved signal, that is, it must have a resolution function associated with it. A guarded signal may be assigned values under the control of a guard expression. It may also be assigned values where a signal GUARD has been explicitly declared. Also, if a guarded signal is assigned a value using a concurrent signal assignment, then the signal assignment must have the keyword **guarded** (a concurrent signal assignment with the **guarded** keyword is called a *guarded assignment*).

A guarded signal behaves differently from other signals in that when the signal GUARD (either explicitly or implicitly declared) gets the value FALSE, the driver to the guarded signal becomes disconnected after a specific time, called the *disconnect time*. On the other hand, in an unguarded signal, if the signal GUARD gets the value FALSE, any new events on the signals appearing in the expression do not influence the value of the target signal; the driver continues to drive the target signal with the old value. To understand this difference better, consider the following block statement, B1.

```
architecture GUARDED_EX of EXAMPLE is
  signal GUARD_SIG: WIRED_OR BIT register;
  signal UNGUARD_SIG: WIRED_AND BIT;
begin
  B1: block ( guard-expression )
  begin
    GUARD_SIG <= guarded expression1 after TIME1;
```

```
        UNGUARD_SIG <= guarded expression2 after TIME2;
    end block B1;
end GUARDED_EX;
```

Transforming the guarded signal assignment statements into their equivalent process statements, the block B1 now looks like:

```
B1: block ( guard-expression )
begin
  process
  begin
    if GUARD then
      GUARD_SIG <= expression1 after TIME1;
    else
      GUARD_SIG <= null;           -- Disconnect driver to GUARD_SIG.
    end if;
    wait on signals-in-expression1, GUARD;
  end process;
  process
  begin
    if GUARD then
      UNGUARD_SIG <= expression2 after TIME2;
    end if;
    wait on signals-in-expression2, GUARD;
  end process;
end block B1;
```

The process statement for the guarded signal GUARD_SIG has an explicit signal assignment statement that disconnects its driver, while there is no such statement for the unguarded signal UNGUARD_SIG. As this example shows, a driver of a guarded signal can be explicitly disconnected by assigning a **null** value to the signal. Such a statement is called a *disconnection statement*.

Let us now explore the differences between a register and a bus signal. A bus signal represents a hardware bus in that when all drivers to the signal become disconnected (as might be the case on a real hardware bus), the value of the signal is determined by calling the resolution function with all the drivers off. A register signal, on the other hand, models a storage component (that may be multiply driven) in that if all drivers to the signal become disconnected, the resolution function is not called and its previous value is retained. With a bus signal, the previous value is lost. Also, bus signals may be either ports of an entity or locally declared signals, whereas register signals can only be locally declared signals.

The disconnect time for a guarded signal can be specified using a *disconnection specification*. The syntax of a disconnection specification is:

```
disconnect guarded-signal-name : signal-type after time-expression ;
```

This is an example of a disconnection specification.

```
disconnect GUARD_SIG: BIT after 8 ns;
```

This implies that the driver of signal GUARD_SIG will get disconnected 8 ns after the corresponding GUARD goes false. This is best understood by looking at how disconnect time affects the disconnection statement that appears in the equivalent process statement of a guarded signal assignment statement. In general, a disconnection specification will modify a disconnection statement to the form:

```
guarded-signal-name <= null after disconnect-time;
  -- Where disconnect time is the time specified in the disconnection
  -- specification for the guarded signal.
```

If no disconnect time is specified for a guarded signal, a default delta delay is assumed. The above disconnection specification for signal GUARD_SIG modifies the equivalent process for the guarded signal assignment to the following.

```
process
begin
  if GUARD then
    GUARD_SIG <= expression1 after TIME1;
  else
    GUARD_SIG <= null after 8 ns;                -- After disconnect time.
  end if;
  wait on signals-in-expression1, GUARD;
end process;
```

The disconnection specification is useful in modeling decay times, for example, capacitance delay on buses. An alternate way of specifying disconnect time is by assigning a value **null** to the signal explicitly using a disconnection statement, as shown next.

```
S1 <= null after 10 ns;
```

This statement specifies that the driver of S1 will be disconnected after 10 ns. Thereafter, this driver does not contribute to the resolved value of the signal. However, such a statement can appear only as a sequential statement, and the target signal must be a guarded signal.

Here is a more comprehensive example.

```
use WORK.RF_PACK.all;
-- Package RF_PACK contains functions WIRED_AND
-- and WIRED_OR.
entity GUARDED_SIGNALS is
  port (CLOCK: in BIT; N: in INTEGER);
end;

architecture EXAMPLE of GUARDED_SIGNALS is
  signal REG_SIG: WIRED_AND INTEGER register;
  signal BUS_SIG: WIRED_OR INTEGER bus;
  disconnect REG_SIG: INTEGER after 50 ns;
  disconnect BUS_SIG: INTEGER after 20 ns;
begin
  BX: block (CLOCK='1' and not CLOCK'STABLE)
  begin
    REG_SIG <= guarded N after 15 ns;
    BUS_SIG <= guarded N after 10 ns;
  end block BX;
end EXAMPLE;
```

On a rising edge on the signal CLOCK, say at time *T*, the current value of N is scheduled to be assigned to signal REG_SIG after 15 ns, that is, at *T*+15 ns, and to signal BUS_SIG after 10 ns, that is, at *T*+10 ns.

However, because of the disconnection specifications, the drivers to signals REG_SIG and BUS_SIG are scheduled to be disconnected after 15+50 ns, that is, at *T*+65 ns, and after 10+20 ns, that is, at *T*+30 ns, respectively. At time *T*+30 ns, the function WIRED_OR is called to determine the value for the signal BUS_SIG, even if all its drivers are off. At time *T*+65 ns, the driver to signal REG_SIG disconnects, the value on signal REG_SIG is retained, and the resolution function WIRED_AND is not called (since there are zero drivers).

The following example describes a 4-by-1 multiplexer using a block statement.

```
library BOOK_LIB;
use BOOK_LIB.UTILS_PKG.WIRED_OR;
entity MUX is
  port (DIN: in BIT_VECTOR(0 to 3); S: in BIT_VECTOR(0 to 1);
        Z: out BIT);
end MUX;

architecture BLOCK_EX of MUX is
  constant MUX_DELAY: TIME := 5 ns;
  signal TP: WIRED_OR BIT bus;
begin
  B1: block (S = "00")
```

```
    begin
      TP <= guarded DIN(0);                    -- Line 12.
    end block B1;
    B2: block (S = "01")
    begin
      TP <= guarded DIN(1);
    end block B2;
    B3: block (S = "10")
    begin
      TP <= guarded DIN(2);
    end block B3;
    B4: block (S = "11")
    begin
      TP <= guarded DIN(3);
    end block B4;
    Z <= TP after MUX_DELAY;
  end BLOCK_EX;
```

Notice that a resolution function is needed for signal TP since it has more than one driver. This function WIRED_OR exists in the package UTILS_PKG that resides in library BOOK_LIB. The signal TP is declared to be a guarded signal of the bus kind. If the value of the select line S is "00", DIN(0) is assigned to signal TP. When the select line has any other value, the driver to signal TP from the guarded assignment on line 12 gets disconnected, and thereafter does not contribute to the effective value of signal TP. Since no disconnect time is specified for signal TP, the default disconnect time is a delta delay. In this example, at any one instant only one driver will always be driving signal TP, which thus models a multiplexer.

10.7 Attributes

An attribute is a value, function, type, range, signal, or constant that can be associated with certain names within a VHDL description. These names could be, among others, an entity name, an architecture name, a label, or a signal. For example, a record that contains the X and Y coordinates of a component placement could be associated with an entity name which describes the placement for that entity in a physical layout; a capacitance value can be associated with all signals of a specific type.

A large number of attributes are predefined in the language. These are described in Section 10.7.2. The language also provides the facility to associate user-defined attributes with names.

10.7.1 User-Defined Attributes

User-defined attributes are constants of any type (except access or file type). They are declared using attribute declarations. An *attribute declaration* declares the name of the attribute and its type; the declaration is of the form:

```
attribute attribute-name : value-type ;
```

For example,

```
type COMP_LOCATION is
  record
    X,Y: INTEGER;
  end record;
type THICKNESS is range 0 to 1000
  units
    micron;
  end units;
type FARADS is range 0 to 5000
  units
    pf;
  end units;
attribute PLACEMENT: COMP_LOCATION;
attribute LENGTH: THICKNESS;
attribute CAPACITANCE: FARADS;
```

These user-defined attributes have not yet been associated with any name. An *attribute specification* is used to associate a user-defined attribute with a name and to assign a value to the attribute. The syntax for an attribute specification is:

```
attribute attribute-name of item-names : name-class is expression ;
```

The *item-names* is a list of one or more names of an entity, architecture, configuration, component, label, signal, variable, constant, type, subtype, package, procedure, or function. The *name-class* indicates the class type, that is, whether it is an entity, architecture, label, or others. The attribute name must have been declared earlier using an attribute declaration. The attribute specification associates an attribute with a group of names that belong to a name class; the expression, whose value must belong to the type of attribute, specifies the value of the attribute. Some examples of attribute specifications are:

```
attribute CAPACITANCE of CLK, RESET: signal is 20 pf;
attribute LENGTH of RX_READY: signal is 3 micron;
```

In the first example, the attribute CAPACITANCE is associated with two signals, CLK and RESET, and the attribute has the value 20 pf.

The item name in the attribute specification can also be replaced with the keyword **all** to indicate all names belonging to that name class. In the following example, the attribute CAPACITANCE is associated with all variables, and the attribute value is set to 0 pf for all variables.

```
attribute CAPACITANCE of all: variable is 0 pf;
```

After having created an attribute and then associated it with a name, the value of the attribute can then be used in an expression by referring to:

```
item-name ' attribute-name        -- The single quote is often read as "tick".
```

For example, RX_READY'LENGTH has the value of 3 micron. Here is a bigger example.

```
architecture PLACEMENT of NAND_GATE is
  component NAND_COMP
    port (IN1, IN2: in BIT; OUT1: out BIT);
  end component;
  type DOUBLE_INT is
    record
      X, Y: INTEGER;
    end record;

  attribute PLACEMENT: DOUBLE_INT;
  attribute SIZE: DOUBLE_INT;
  attribute PLACEMENT of N1: label is (50, 45);
  attribute SIZE of N1: label is (2, 4);
  signal PERIMETER: INTEGER;
  signal A, B, Z: BIT;
begin
  N1: NAND_COMP port map (A, B, Z);
  PERIMETER <= 2 * (N1'SIZE.X + N1'SIZE.Y);
end NAND_PLACE;
```

When an attribute is associated with an overloaded subprogram or an overloaded enumeration literal, a signature needs to be specified. The signature, when used in an attribute specification or attribute name, uniquely identifies the subprogram or literal. Here are examples of using signatures with attributes.

```
function TO_STDLOGICVECTOR
  (OPD: INTEGER; SIZE: NATURAL) return STD_LOGIC_VECTOR;
function TO_STDLOGICVECTOR (OPD: BIT_VECTOR;
  SIZE: NATURAL) return STD_LOGIC_VECTOR;

attribute BUILT_IN: BOOLEAN;

attribute BUILT_IN of TO_STDLOGICVECTOR [ INTEGER,
  NATURAL return STD_LOGIC_VECTOR ] : function is TRUE;
attribute BUILT_IN of TO_STDLOGICVECTOR [ BIT_VECTOR,
  NATURAL return STD_LOGIC_VECTOR ] : function is FALSE;
```

Two overloaded functions are declared. The attribute declaration declares the attribute BUILT_IN to be of type BOOLEAN. The two attribute specifications associate the attribute BUILT_IN with the two corresponding overloaded functions and assigns a value to the attribute as well. Note the use of signatures to distinguish between the two overloaded functions.

The value of the attribute for the first overloaded function can then be accessed as:

```
TO_STDLOGICVECTOR
  [ INTEGER, NATURAL return STD_LOGIC_VECTOR ] 'BUILT_IN
```

which has the value TRUE. Note that, again, a signature has been used in the attribute name to distinguish between the two overloaded functions.

User-defined attributes are useful for annotating VHDL models with tool-specific information.

10.7.2 Predefined Attributes

There are five classes of predefined attributes:

1. Value attributes: These return a constant value.
2. Function attributes: These call a function that returns a value.
3. Signal attributes: These create a new signal.
4. Type attributes: These return a type name.
5. Range attributes: These return a range.

Value Attributes

If T is any scalar type or subtype:

- T'LEFT returns the left bound, that is, the leftmost value of T.
- T'RIGHT returns the right bound, that is, the rightmost value of T.
- T'HIGH returns the upper bound, that is, the largest value in T.
- T'LOW returns the lower bound, that is, the smallest value in T.
- T'ASCENDING returns TRUE if the type has an ascending range; otherwise, it returns FALSE.

For example, if:

```
type ALLOWED_VALUE is range 31 downto 0;
type WEEK_DAY is (SUN, MON, TUE, WED, THU, FRI, SAT);
  -- SUN is the smallest value and SAT is the largest value
  -- of type WEEK_DAY.
subtype WORK_DAY is WEEK_DAY range FRI downto MON;
```

then the following equivalence relations are true.

```
ALLOWED_VALUE'LEFT = 31
ALLOWED_VALUE'HIGH = 31

ALLOWED_VALUE'RIGHT = 0
ALLOWED_VALUE'LOW = ALLOWED_VALUE'RIGHT

WEEK_DAY'LEFT = SUN;
WEEK_DAY'LOW = WEEK_DAY'LEFT
WEEK_DAY'RIGHT = SAT;
WEEK_DAY'HIGH = WEEK_DAY'RIGHT

WORK_DAY'RIGHT = MON;
WORK_DAY'LOW = WORK_DAY'RIGHT
WORK_DAY'LEFT = FRI;
WORK_DAY'HIGH = WORK_DAY'LEFT

ALLOWED_VALUE'ASCENDING          is FALSE
WORK_DAY'ASCENDING               is FALSE
WEEK_DAY'ASCENDING               is TRUE
```

If A is a constrained array object, then:

- A'LENGTH(N) returns the number of elements in the Nth dimension (N=1 if not specified).

- A'ASCENDING(N) returns TRUE if the Nth index range of A has an ascending range, else it is FALSE (N=1 if not specified).

For example, if:

```
signal TX_BUS: STD_LOGIC_VECTOR (7 downto 0);
```

then:

```
TX_BUS'LENGTH                is 8
TX_BUS'ASCENDING             is FALSE
```

If E is any name, then:

- E'SIMPLE_NAME is the string representation of the name. All characters in the string are in lower case, except when E is an extended identifier, the case of the identifier is preserved.

Some examples are:

```
RIB'SIMPLE_NAME                      is "rib"
BIT'SIMPLE_NAME                      is "bit"
\7400TTL\'SIMPLE_NAME                is "\7400TTL\"

report "Signal " & TX_BUS'SIMPLE_NAME & " has value 4F.";
-- Generates the output:
        ** Note: Signal tx_bus has value 4F.
```

If E is any name other than a local port or a local generic of a component declaration:

- E'INSTANCE_NAME is a string that contains the hierarchical path starting at the top of the design hierarchy to the name E. Again, all characters in string are in lower case; except when E is an extended identifier, the case of the identifier is preserved. The path contains the name of the instantiated entity-architecture pairs as well.
- E'PATH_NAME is a string that also contains the hierarchical path starting from the top of the design hierarchy to the name E; but in this case, it does not contain the names of the instantiated entity-architecture pairs. All characters in the string are again in lower case, except when E is an extended identifier, the case of the identifier is preserved.

Some examples are:

```
-- A package LOGIC_ARITH in design library DZX:
LOGIC_ARITH'PATH_NAME                  is ":dzx:logic_arith:"
LOGIC_ARITH'INSTANCE_NAME              is ":dzx:logic_arith:"

-- A function MAX in package LOGIC_ARITH in design library DZX:
MAX'PATH_NAME                          is ":dzx:logic_arith:max"
MAX'INSTANCE_NAME                      is ":dzx:logic_arith:max"

-- A variable SUP declared in a process RX in an architecture
-- body RX_DW of an entity RX_TOP:
SUP'PATH_NAME                          is ":rx_top:rx:sup"
SUP'INSTANCE_NAME                      is ":rx_top(rx_dw):rx:sup"

-- A type COLORS declared in an unlabeled process in an architecture
-- body SM of entity DCW:
COLORS'PATH_NAME                       is ":dcw: : colors"
COLORS'INSTANCE_NAME                   is ":dcw (sm) : : colors"

--Variable TEMP declared in an unlabeled process in an architecture
-- body XOR2_ARCH of an entity XOR2 that is instantiated as a
-- component with label LAB which occurs in a block labeled
-- BAR that is in an architecture body FA_STR of entity FA:
TEMP'PATH_NAME                         is ":fa:bar:lab::temp"
TEMP'INSTANCE_NAME                     is
                    ":fa(fa_str):bar:lab@xor2(xor2_arch)::temp"
```

'PATH_NAME and 'INSTANCE_NAME are identical for items declared inside a package.

Function Attributes

These attributes represent functions that are called to obtain a value. They are often used to convert values from an enumeration or physical type to an integer type. If T is a discrete type, a physical type, or a subtype, then:

- T'POS(V) returns the position number of the value V in the ordered list of values of T.
- T'VAL(P) returns the value of the type that corresponds to position P.
- T'SUCC(V) returns the value of the parameter whose position is one larger than the position of value V in T.
- T'PRED(V) returns the value of the parameter whose position is one less than the position of value V in T.
- T'LEFTOF(V) returns the value of the parameter that is to the left of value V in T.

- T'RIGHTOF(V) returns the value of the parameter that is to the right of value V in T.

For ascending ranges,

```
T'SUCC(X) = T'RIGHTOF(X)        -- for X /= T'RIGHT.
T'PRED(X) = T'LEFTOF(X)         -- for X /= T'LEFT.
```

For descending ranges,

```
T'SUCC(X) = T'LEFTOF(X)         -- for X /= T'LEFT.
T'PRED(X) = T'RIGHTOF(X)        -- for X /= T'RIGHT.
```

For example, if:

```
type STATUS is (SILENT, SEND, RECEIVE);
subtype DELAY_TIME is TIME range 50 ns downto 10 ns;
```

then the following equivalence relations are true.

```
STATUS'POS(SEND) = 1
STATUS'VAL(2) = RECEIVE
DELAY_TIME'SUCC(21 ns) = 21 ns + 1 fs
DELAY_TIME'PRED(10 ns) is an error
DELAY_TIME'LEFTOF(29 ns) = 29 ns + 1 fs
DELAY_TIME'SUCC(29 ns) = DELAY_TIME'LEFTOF(29 ns)
DELAY_TIME'RIGHTOF(11 ns) = 11 ns – 1 fs
DELAY_TIME'PRED(11 ns) = DELAY_TIME'RIGHTOF(11 ns)
STATUS'SUCC(RECEIVE) will cause an error
STATUS'PRED(RECEIVE) = SEND
STATUS'LEFTOF(RECEIVE) = STATUS'PRED(RECEIVE)
STATUS'SUCC(SILENT) = SEND
STATUS'RIGHTOF(SILENT) = STATUS'SUCC(SILENT)
```

Here are some more examples.

```
CHARACTER'VAL(65)               is the character 'A'
CHARACTER'POS('A')              is 65
-- Type CHARACTER is defined in package STANDARD.
```

If A is a constrained array object, then:

- A'LEFT(N) returns the left bound of the Nth dimension of the array.
- A'RIGHT(N) returns the right bound of the Nth dimension.
- A'LOW(N) returns the lower bound of the Nth dimension.
- A'HIGH(N) returns the upper bound of the Nth dimension.

For ascending ranges,

```
A'LEFT(N) = A'LOW(N)
A'RIGHT(N) = A'HIGH(N)
```

For descending ranges,

```
A'LEFT(N) = A'HIGH(N)
A'RIGHT(N) = A'LOW(N)
```

In all the previous cases, N=1 if not specified. Some more examples are shown next.

If:

```
type COST_TYPE is array (7 downto 0, 0 to 3) of INTEGER;
variable COST_MATRIX: COST_TYPE;
```

then the following equivalence relations are true.

```
COST_MATRIX'LEFT(2) = 0
COST_MATRIX'LOW(1) = 0
COST_MATRIX'RIGHT(2) = 3
COST_MATRIX'HIGH(1) = 7
```

If S is a signal object, then:

- S'EVENT returns true if an event occurred on signal S in the current delta.
- S'ACTIVE returns true if signal S is active in the current delta.

  It is important to understand the difference between an event on a signal and a signal being active. A signal is said to be active when a new value is assigned to the signal, even if the new value is the same as the old value; an event is said to have occurred only if the new value is different from the old value.

- S'LAST_EVENT returns the time elapsed since the last event on signal S.
- S'LAST_ACTIVE returns the time elapsed since the last time signal was active.
- S'LAST_VALUE returns the value of S, before the last event.
- S'DRIVING returns the value FALSE if, in the enclosing process, the driver for signal S is disconnected. Otherwise, it returns TRUE.

- S'DRIVING_VALUE returns the current value of the driver for S in the process in which this attribute is used. It is illegal to access this attribute value when S'DRIVING is false.

The following examples show their usage in expressions.

```
signal CLOCK: BIT;
constant SETUP_TIME: TIME := 5 ns;
signal A: BIT;
signal COUNT: INTEGER;

CLOCK = '1' and CLOCK'EVENT          -- Denotes a rising edge on
                                     -- signal CLOCK.
A'LAST_EVENT < SETUP_TIME
COUNT = 20 and COUNT'LAST_VALUE = 10
```

Given the following process statement,

```
process
begin
  DRD <= 'Z', '1' after 5 ns, null after 15 ns, 'U' after 25 ns;
  . . . DRD'DRIVING . . .           -- Use of 'DRIVING attribute.
  . . . DRD'DRIVING_VALUE . . .     -- Use of 'DRIVING_VALUE
                                    -- attribute.
  wait;
end process;
```

Figure 10-3 shows the values of these two attributes used in the process.

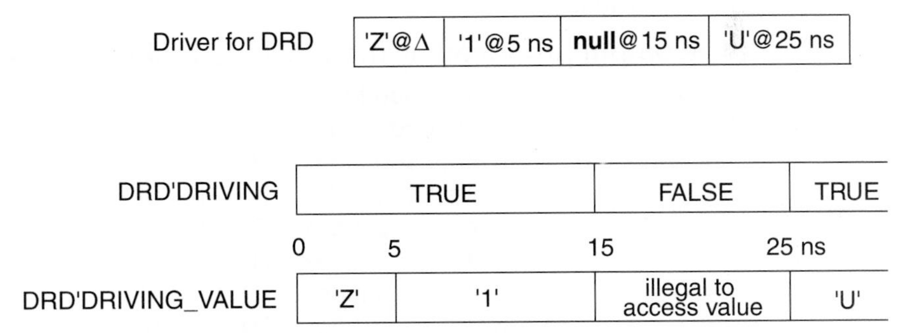

Figure 10-3 'DRIVING and 'DRIVING_VALUE attribute example.

If T is any scalar type or a subtype, then:

- T'IMAGE (X) returns the string representation of the value X that is of type T. All characters are in lower case; except when X is an escaped identifier, the case of the identifier is preserved.
- T'VALUE (X) returns the value of T whose string representation is X.

Given:

```
type ALLOWED_VALUE is range 31 downto 0;
type WEEK_DAY is (SUN, MON, TUE, WED, THU, FRI, SAT);
subtype WORK_DAY is WEEK_DAY range FRI downto MON;
```

then:

```
ALLOWED_VALUE'IMAGE(31)          is "31"
WEEK_DAY'IMAGE (SUN)             is "sun"
WORK_DAY'IMAGE (FRI)             is "fri"

ALLOWED_VALUE'VALUE ("16")       is 16
WEEK_DAY'VALUE ("sat")           is SAT
WORK_DAY'VALUE ("mon")           is MON
```

An example of using the 'IMAGE attribute to convert an integer to a string and a BIT to a string is shown next.

```
constant RX: BIT_VECTOR(0 to 7) := "11001110";
. . .
for J in RX'RANGE loop
  report "Loop index = " & "RX(" & INTEGER'IMAGE(J) & ") = "
         & BIT'IMAGE(RX(J));
end loop;
```

The output produced by the `for` loop looks like this.

```
** Note: Loop index = RX(0) = '1'
** Note: Loop index = RX(1) = '1'
** Note: Loop index = RX(2) = '0'
** Note: Loop index = RX(3) = '0'
** Note: Loop index = RX(4) = '1'
** Note: Loop index = RX(5) = '1'
** Note: Loop index = RX(6) = '1'
** Note: Loop index = RX(7) = '0'
```

Here is an example of 'IMAGE that prints time and a value of type STD_LOGIC.

```
variable FIRST_FLAG: STD_LOGIC := '1';
. . .
```

```
report "Value of FIRST_FLAG = " &
    STD_LOGIC'IMAGE(FIRST_FLAG) & " at time " &
    TIME'IMAGE(NOW);
```

Execution of this statement produces:

```
** Note: Value of FIRST_FLAG = '1' at time 0 ns
```

Signal Attributes

These attributes create new signals from the signals with which they are associated. These attributes therefore create implicit signals, as compared to explicit signals that are created using signal declarations. If S is a signal object, then:

- S'DELAYED (T) is a new signal that is the same type as signal S but delayed from S by time T. If T is not specified, a delta delay is assumed.
- S'STABLE (T) is a Boolean signal that is true when signal S has not had any event for time T. If T is not specified, it implies current delta.
- S'QUIET (T) creates a Boolean signal that is true when S has not been active for time T. If T is not specified, it implies a current delta.
- S'TRANSACTION creates a signal of type BIT that toggles its value every time signal S becomes active.

Assuming:

```
signal CLOCK_SKEW, CTRL_A: BIT;
```

Figure 10-4 shows some examples of signal attributes.

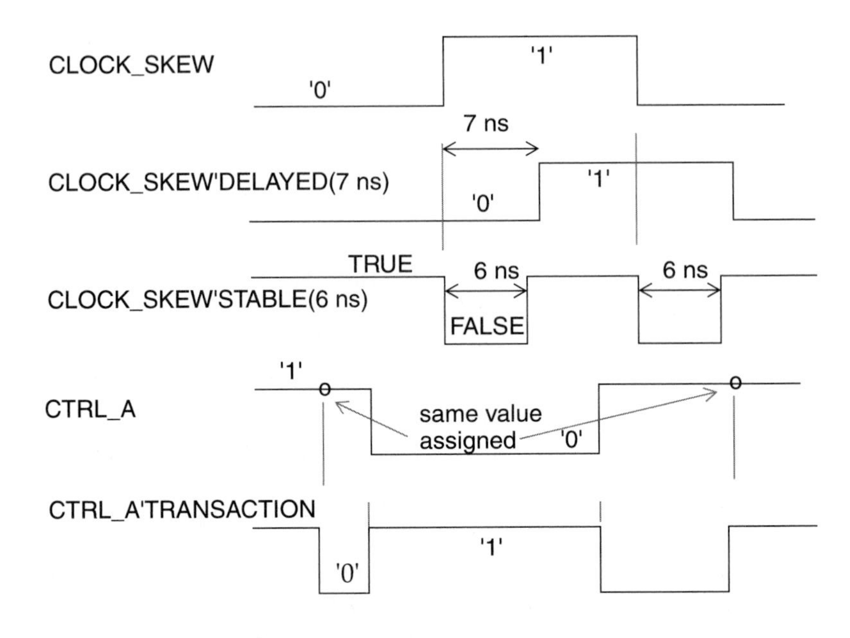

Figure 10-4 Signal attribute examples.

Here is an example of use of the 'STABLE attribute to check for a setup constraint. It is required to report a violation if signal DATA has not been stable for the specified SETUP time before the falling edge of the signal CLK occurs.

```
process
  constant SETUP: TIME := 1.2 ns;
begin
  wait until CLK = '0';
  assert DATA'STABLE(SETUP)
    report "Setup violation"
    severity WARNING;
end process;
```

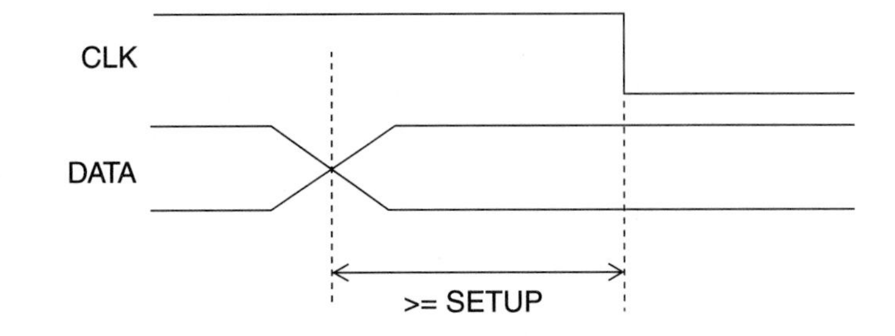

Figure 10-5 Setup constraint check.

Figure 10-5 shows what the process is checking for. The `wait` statement waits for a negative edge to occur on signal CLK. The execution of the next statement checks to see if DATA has been stable for at least the SETUP time, if not, a message "Setup violation" is reported.

It is important to understand the difference between function attributes for signals and signal attributes. Signal attributes create new signals and therefore cause events when used in concurrent statements, whereas function attributes for signals do not create new signals and therefore cannot create new events. Signal attributes can therefore be used wherever a signal is expected; for example, in the sensitivity list of a process statement or as a signal parameter in a procedure. For example,

```
PSTABLE: process (A, B, CLK'STABLE)
begin
  . . .
end process;
```

It is illegal to use CLK'EVENT in the sensitivity list for the above process since only signals are allowed in the sensitivity list of a process. Therefore, function attributes of signals should not be used in expressions where the intention is to create new events.

Here is another example that highlights the difference between the 'EVENT attribute and the 'STABLE attribute.

```
B1: block (CLK = '1' and CLK'EVENT)
begin
  Q <= guarded D;
end block;
```

In this example, the value of the guard expression is independent of CLK'EVENT since this is a function and never changes it's value. Thus the block statement behaves like a level-sensitive assignment: whenever CLK is '1', value of D is passed to Q. If the intention of the block statement was to model an edge-triggered assignment, the 'STABLE attribute must be used as shown in the following block statement.

```
B2: block (CLK = '1' and not CLK'STABLE)
begin
  Q <= guarded D;
end block;
```

CLK'STABLE is a signal; so any transition on CLK is recognized by the expression "**not** CLK'STABLE" and the guard expression indicates a rising-clock edge. Value of D is transferred to Q only on the rising edge of CLK.

Type Attributes

If T is any type or subtype, T'BASE, which is the only type attribute, returns the base type of T. This attribute cannot be used in expressions as such since it returns a base type, but it can be used in conjunction with other attributes. For example, if:

```
type ALU_OPS is (ADDOP, SUBOP, MULOP, DIVOP, ANDOP,
                 NANDOP, OROP, NOROP);
subtype ARITH_OPS is ALU_OPS range ADDOP to DIVOP;
```

then:

```
ARITH_OPS'BASE            is ALU_OPS
```

which is the base type, and it can be used in an expression such as:

```
ARITH_OPS'BASE'LEFT
```

which has the value ADDOP.

Range Attributes

If A is a constrained array object, then:

- A'RANGE (N) returns the Nth index range of A (N=1 if not specified).

- A'REVERSE_RANGE(N) returns the Nth index range reversed (N=1 if not specified).

For example, if:

```
variable WBUS: STD_LOGIC_VECTOR(7 downto 0);
```

then:

WBUS'RANGE returns the range "7 **downto** 0" (quotations not included)

while:

WBUS'REVERSE_RANGE returns the range "0 **to** 7".

The range attributes provide a mechanism for specifying the index constraint in **for** loops and for-generate statements in a parameterized way. For example, a use of this attribute in a **for** loop is shown next.

```
for INDEX in WBUS'REVERSE_RANGE loop
  ...
end loop;
```

'FOREIGN attribute

The 'FOREIGN attribute is a user-defined attribute, predefined in package STANDARD, that allows a subprogram or an architecture body to have non-VHDL implementations. The string value appearing in the attribute specification for the 'FOREIGN attribute contains implementation-dependent information. The algorithm performed by such a subprogram or an architecture body that is decorated with the 'FOREIGN attribute is implementation-defined. The general form of the attribute specification for the 'FOREIGN attribute is:

```
attribute FOREIGN of architecture-or-subprogram-name :
  [ architecture | function | procedure ]
  is "implementation-dependent-information";
```

Some examples follow.

```
entity AND2 is
  port (A, B: in BIT; C: out BIT);
end AND2;

architecture C_MODEL of AND2 is
  attribute FOREIGN of C_MODEL: architecture
    is "xxand2(A, B, C)";
begin
```

```
    -- No concurrent statements necessary here, since
    -- they will never be executed.
end C_MODEL;

impure function RANDOM (SEED: REAL) return REAL;
attribute FOREIGN of RANDOM: function is "rand(seed)";
```

The presence of an attribute specification for the 'FOREIGN attribute in the architecture body implies that when the entity AND2 with architecture C_MODEL is invoked during simulation, implementation-defined actions occur; in this example, the system function "xxand2" is called. Therefore, no other statements or declarations need be present within the architecture body.

The function RANDOM also has the 'FOREIGN attribute associated with it. Such a subprogram is called a *foreign subprogram*. Whenever function RANDOM is called, implementation-defined actions occur; in this example, a call to the system function "rand" is performed. In this case, there is no need for a subprogram body. Alternately, the attribute specification for the 'FOREIGN attribute may be present in the subprogram body; in this case, the subprogram body need not contain any other statements or declarations.

Parameter-passing mechanisms are not defined by the language for foreign subprograms.

10.8 Aggregate Targets

It is possible for the target of a variable assignment statement or a signal assignment statement to be an aggregate. An aggregate target represents a combination of one or more names, that is, of an array type or a record type.

Here is an example.

```
signal A, B, C: BIT;

(A, B, C) <= BIT_VECTOR'("100");
```

The left-hand side represents an aggregate target made up of three individual signals. The signal assignment statement indicates that signal A is assigned the value '1', and signals B and C are assigned the value '0'. Here is another example.

```
type COUNT_REC is
   record
      INPUTS, OUTPUTS: POSITIVE;
   end record;
variable BOX1, BOX2: COUNT_REC;
variable TOTAL_IN, TOTAL_OUT: POSITIVE;
function "+" (A1, A2: COUNT_REC) return COUNT_REC;

(TOTAL_IN, TOTAL_OUT) := BOX1 + BOX2;
```

In this case, variable TOTAL_IN gets the value of the INPUTS element of the overloaded function result, and TOTAL_OUT gets the value of the OUTPUTS element of the function result.

In general, an assignment to an aggregate target causes subelement values in the right-hand-side expression to be assigned to the corresponding subelements in the left-hand-side target.

10.9 More on Block Statements

Two uses of a block statement were described in Chapter 5. One other use of a block statement is in representing a portion of a design. This means that an architecture body can be made up of a number of block statements, each representing a portion of a design. Block statements, in turn, can contain one or more concurrent statements, including other block statements. Block statements thus provide a mechanism for creating a hierarchy within an architecture body.

When a portion of a design is represented using a block statement, it is possible to use arbitrary constant and signal names within the block and then to associate these names to those outside of that block by using the generic map and port map in the block header. The next example shows a description of a full-adder that has been partitioned into three blocks (where each could have been designed by a separate designer). Two of these blocks, HA1 and HA2, represent the two half-adders, and the block OR1 represents an `or` gate. Different styles of modeling are chosen in describing the half-adders and the `or` gate to show that block statements can contain any form of concurrent statement.

```
entity FULL_ADDER is
  generic (HA_DELAY: TIME := 5 ns);
  port (A, B, CIN: in BIT; SUM, COUT: out BIT);
end FULL_ADDER;

architecture BLOCK_VIEW of FULL_ADDER is
  signal S1, C1, C2: BIT;
begin
  HA1: block
    generic (CARRY_DELAY: TIME);
    generic map (CARRY_DELAY => HA_DELAY);
    port (IN1, IN2: in BIT; SUM, CARRY: out BIT);
    port map (IN1 => A, IN2 => B, SUM => S1, CARRY => C1);
  begin
    SUM <= (IN1 and not IN2) or (IN2 and not IN1);
    CARRY <= IN1 and IN2 after CARRY_DELAY;
  end block HA1;

  HA2: block
    port (X, Y: in BIT; S, C: out BIT);
    port map (X => CIN, Y => S1, S => SUM, C => C2);
  begin
    process (X, Y)
    begin
      S <= X xor Y;
      C <= X and Y after HA_DELAY;
    end process;
  end block HA2;

  OR1: block
    port (A, B: in BIT; C: out BIT);
    port map (A => C1, C => COUT, B => C2);
    component OR_GATE
      port (X, Y: in BIT; Z: out BIT);
    end component;
  begin
    O1: OR_GATE port map (A, B, C);
  end block OR1;
end BLOCK_VIEW;
```

Block HA1 describes a half-adder using concurrent signal assignment statements. The generic and port maps associate local names (those within the block) to external names (names outside the block). Signal A, which is an input to the entity FULL_ADDER, is associated with signal IN1 used inside block HA1, signal S1 defined in the architecture body is associated with signal SUM declared within the HA1 block, and so on. Block HA2 describes a half-adder using a process statement, while block OR1 contains a component instantiation statement to represent an `or` gate. Port signals declared within each block are local to that block, while signals declared in the architecture body and in the entity declaration are global to all the blocks. For example, signal SUM in HA1 refers to the signal SUM declared in that block and not to the output signal SUM. The HA_DELAY used in block HA2 is the generic declared in the entity declaration.

When components are instantiated inside a block statement, it is necessary to specify a block configuration for the block when writing the configuration declaration for such a design. Here is an example of a configuration declaration for the previous FULL_ADDER entity.

```
library CMOS_LIB;
configuration BLOCK_VIEW_CON of FULL_ADDER is
  for BLOCK_VIEW
    for OR1                                  -- A block configuration.
      for O1: OR_GATE
        use entity CMOS_LIB.OR2(OR2);
      end for;
    end for;
  end for;
end BLOCK_VIEW_CON;
```

Within the block configuration for the architecture body, there is one block configuration, that for block OR1. Blocks HA1 and HA2 do not have any component instantiations and therefore have nothing to bind. It is not necessary to specify the block configurations for such blocks. The block OR1 has one component that is bound to an entity represented by the entity-architecture pair from library CMOS_LIB.

Since a block statement is just another concurrent statement, block statements can be nested. Here is an example.

```
B1: block
  signal A, B: BIT;
begin
  . . .
  B2: block
    signal B, C: BIT;
  begin
    A <= B and C;                   -- First signal assignment.
    C <= B1.B;                      -- Second signal assignment.
  end block B2;
end block B1;
```

The signal B used in the first signal assignment in block B2 refers to the signal B declared in that block. If the signal B declared in block B1 needs to be used within block B2, the signal must be explicitly qualified by the block label, as shown in the second statement in block B2.

10.10 Shared Variables

A variable that is declared outside of a process or a subprogram is called a *shared variable*. A shared variable can be read and updated by more than one process.

The interpretation of shared variables is not defined by the language; therefore, their usage is not recommended.

10.11 Groups

It is sometimes useful to group certain items so that some common properties can be associated with the group. For example, it might be useful to identify groups of two signals so that their path delay could be attributed to the group name. Another example is to identify a group of component labels that are to be attributed with the property of "do not touch", which is a flag indicating blocks not to be touched during logic optimization.

This grouping is achieved by using both of the following:

1. Group template declaration
2. Group declaration

A *group template declaration* declares a template for a group; that is, it specifies the class of items and the number of items that are to appear in a group. Here are some examples of group template declarations.

```
group PATH_DELAY is (signal, signal);
group OP_BINDING is (function, label);
group USED_IN is (signal, label <>);
group RTL_BINDING is (label <>);
```

```
group ZERO is (literal <>);
group BUILT_IN is (function <>);
```

PATH_DELAY is a group template that contains exactly two signals. OP_BINDING is a group template that contains one function name and one label. USED_IN is a group template that contains one signal and zero or more labels. RTL_BINDING is a group template that contains zero or more labels. ZERO is a group template with zero or more literals. BUILT_IN is a group template with zero or more function names.

A *group declaration* declares the name of the group and also the named items that are part of the group. Here are examples of group declarations.

```
group ALU1: RTL_BINDING (PLUS1, PLUS2, PLUS3);
group ALU2: RTL_BINDING (SUB1, PLUS4);
group P1: PATH_DELAY (CLK, SDN);
group B1: BUILT_IN (TO_STDLOGIC, TO_STDLOGICVECTOR,
                    TO_INTEGER);
group Z1: ZERO (STD.STANDARD.'0', ATT.ATT_MVL.'0',
                IEEE.STD_LOGIC_1164.'0');
```

ALU1 is the name of a group that is of the form of the group template RTL_BINDING. The elements in this group are the labels PLUS1, PLUS2, PLUS3. ALU2 is another group of the same template, RTL_BINDING. In this case, the group consists of two labels, SUB1 and PLUS4. Z1 is the group that is of the template form ZERO, and it has the three specified elements in the group.

10.12 More on Ports

This section elaborates on the key differences between **in**, **out**, **buffer** and **inout** ports. Port behavior differs on how they behave internal to the entity and on how they behave external to the entity, especially in what other kind of ports can be connected. Figure 10-6 shows pictorially the various port behaviors internal to an entity and external to the entity.

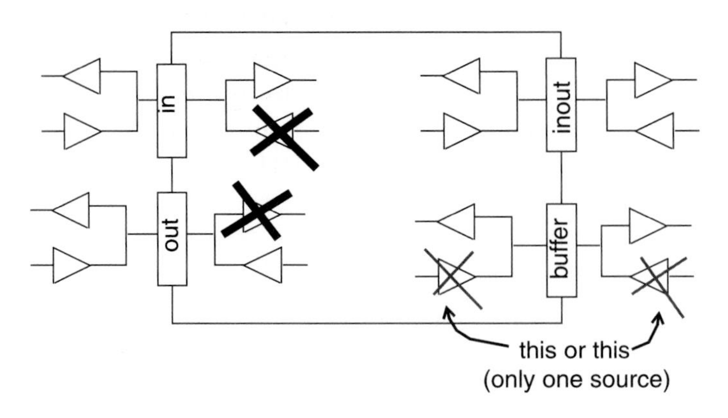

Figure 10-6 Port modes: **in**, **out**, **inout**, **buffer**.

1. Internal to entity:
   - **in** port: Can only read a value. A value can never be assigned to it.
   - **out** port: Cannot be read. It can be assigned a value, and can have multiple drivers.
   - **inout** port: Can be read. It can be assigned a value, and can have multiple drivers.
   - **buffer** port: Can be read. It can be assigned a value, but can have at most one driver on the entire net connected to the buffer port (either internal to entity or external to entity).
2. External to entity:
   Net connection rules:
   - **in**, **inout**, and **out** port: Can be connected to any net with or without multiple drivers.
   - **buffer** port: Can only be connected to a net that has at most one driver (either internal or external to entity).

   Port connection rules:
   There are rules on what kind of ports (those of parent entity) can be directly connected to the ports of a child entity. This is pictorially shown in Figure 10-7.
   : **in** port: Can be connected to a parent entity's port of mode **in**, **inout** or **buffer**.

: **out** port: Can only be connected to a parent entity's port of mode **out** and **inout**.
: **inout** port: Can only be connected to **inout**.
: **buffer** port: Can only be connected to another port of mode **buffer**.

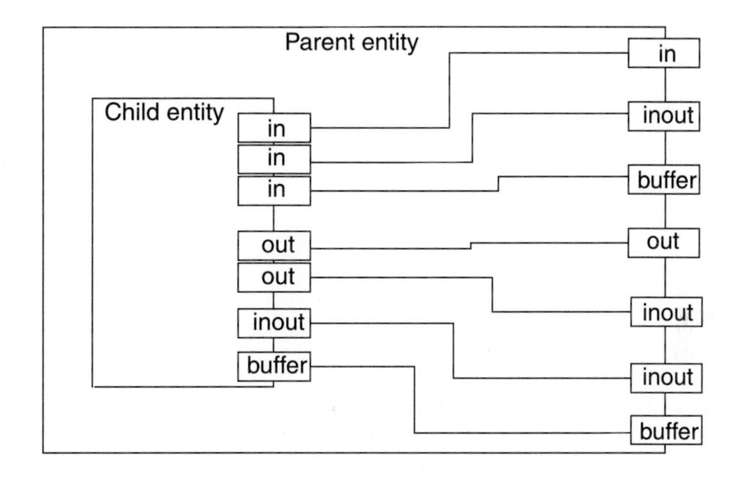

Figure 10-7 Port-to-port connection rules.

For example, an **in** port can be connected to an **inout** port but an **out** port of child entity cannot be connected to a **buffer** port of the parent. Also notice that a **buffer** port can only be connected to another **buffer** port; it cannot be connected to an **inout** or an **out** port. The purpose of this rule is to ensure that the effective and driving values of the entire net that the buffer port encompasses is always the same. A side effect of this rule is that a buffer port need never be resolved since multiple drivers can never exist on a net connecting a buffer port.

❑

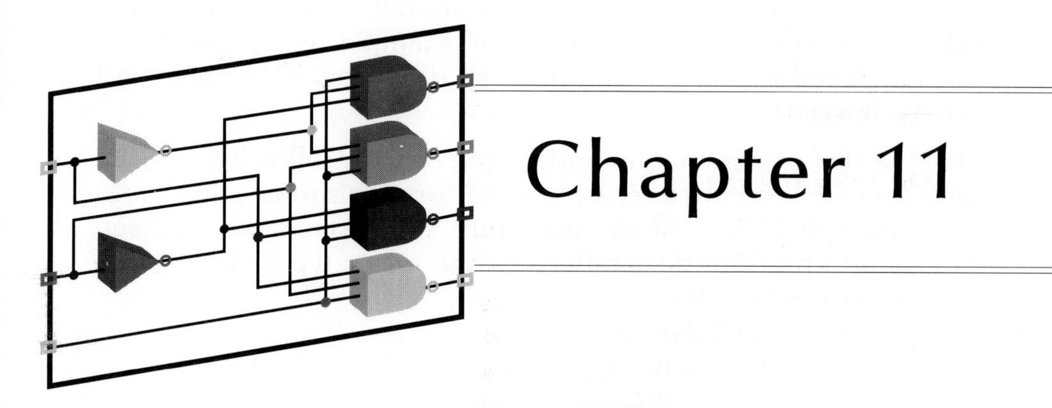

Chapter 11

Model Simulation

This chapter describes the environment necessary for simulating hardware models written in VHDL. These models can be tested by writing test bench models, which can themselves be described in VHDL. Some approaches to writing test bench models are also described in this chapter.

11.1 Simulation

Before beginning to model hardware, the first thing to decide is what values will be used during simulation. Will all objects take values '0' or '1' only, or take values '0', '1', 'U', or 'Z', or use a 46-value logic? The number of values used during simulation will define the basic types that will be used to model hardware. This important decision has to be made prior to writing any model, since the predefined types that the language provides are not sufficient to model values such as 'U' (undefined) and 'Z' (high-impedance). The language provides the capability for creating user-defined types that can be used to model the values required for a particular simulation.

In certain cases, it may be sufficient to model hardware using the predefined types of the language, for example, when modeling hardware at an abstract level where all data may be represented strictly as integers.

For the rest of this chapter and for the chapter following this, we shall assume that we are interested in performing a nine-value simulation. For this, we shall use the IEEE standard package STD_LOGIC_1164. The declarations of the major types declared in this package are:

```
type STD_ULOGIC is (
          'U',      -- Uninitialized
          'X',      -- Forcing unknown
          '0',      -- Forcing 0
          '1',      -- Forcing 1
          'Z',      -- High impedance
          'W',      -- Weak unknown
          'L',      -- Weak 0
          'H',      -- Weak 1
          '-'       -- Don't care
          );
subtype STD_LOGIC is RESOLVED STD_ULOGIC;
  -- RESOLVED is the name of a resolution function.
type STD_ULOGIC_VECTOR is array (NATURAL range <>)
  of STD_ULOGIC;
type STD_LOGIC_VECTOR is array (NATURAL range <>)
  of STD_LOGIC;
```

The leftmost value of type STD_ULOGIC is defined to be a 'U' so that all objects that are defined to be of this base type (also of subtype STD_LOGIC) will have an initial value of 'U' at the start of simulation. In rest of the examples in this book, we shall be using type STD_LOGIC for scalars and type STD_LOGIC_VECTOR for vectors.

The complete package STD_LOGIC_1164 is listed in Appendix E. This package is always compiled into design library IEEE. The content of the package STD_LOGIC_1164 can be accessed by using the `library` and `use` clauses, such as:

```
library IEEE;
use IEEE.STD_LOGIC_1164.all;
```

If a purely behavioral model is being simulated, the previous package may be sufficient. If structural models are to be simulated that reference primitive components defined in a certain library, it is useful to provide a package containing the component declarations for all components that reside in the library. There might be a separate package provided for each class of components, for example, a package that contains the component declarations for CMOS components, a package for all TTL 7400-series components, a package for ECL components, and so on. This would eliminate the need for writing component declarations in every structural description; all that would be necessary is to reference the component declarations package by using a `use` clause. Here is a template for such a package.

```
library IEEE;
use IEEE.STD_LOGIC_1164.all;
package CMOS_COMP is
  component AOI21
    port (A1, A2, B: in STD_LOGIC; Z: out STD_LOGIC);
  end component;
  component FD1S3AX
    port (D, CKA: in STD_LOGIC; Q, QN: out STD_LOGIC);
  end component;
  component INRB
    port (A: in STD_LOGIC; Z: out STD_LOGIC);
  end component;
  -- Other components would be defined similarly.
end CMOS_COMP;
```

This package has to be compiled into a design library, say, CMOSLIB.

There is yet another piece of information that is required before a structural model can be simulated. These are the behavioral models for any primitive components used in the structural model. If the primitive components as defined in the CMOS_COMP package are used, it is necessary to provide behavioral models for these primitive components. Three such examples are shown next.

```
library IEEE;
use IEEE.STD_LOGIC_1164.all;
entity AOI21 is
  port (A1, A2, B: in STD_LOGIC; Z: out STD_LOGIC);
end AOI21;

architecture BEH_MODEL of AOI21 is
begin
  process (A1, A2, B)
    variable TEMP: STD_LOGIC;
  begin
    TEMP := A1 and A2;
    Z <= TEMP nor B;
```

```
    end process;
end BEH_MODEL;

library IEEE;
use IEEE.STD_LOGIC_1164.all;
entity FD1S3AX is
    port (D, CKA: in STD_LOGIC; Q, QN: out STD_LOGIC);
end FD1S3AX;

architecture BEH_MODEL of FD1S3AX is
begin
    process (D, CKA)
        variable TEMP: STD_LOGIC;
    begin
        if CKA = '1' and CKA'EVENT then
            TEMP := D;
        end if;
        Q <= TEMP;
        QN <= not TEMP;
    end process;
end BEH_MODEL;

library IEEE;
use IEEE.STD_LOGIC_1164.all;
entity INRB is
    port (A: in STD_LOGIC; Z: out STD_LOGIC);
end INRB;

architecture BEH_MODEL of INRB is
begin
    Z <= not A;
end BEH_MODEL;
```

The behavioral descriptions for all the primitive library components also need to be compiled into a design library, say, CMOSLIB. At this point, we are ready to begin modeling and simulating hardware.

11.2 Writing a Test Bench

A test bench is a model that is used to exercise and verify the correctness of a hardware model. The expressive power of the VHDL language provides us with the capability of writing test bench models in the same language. A test bench has three main purposes:

1. To *generate* stimulus for simulation (waveforms)
2. To *apply* this stimulus to the entity under test and collect the output responses
3. To *compare* output responses with expected values

The language provides a large number of ways to write a test bench. In this section, we explore only some of these. A typical test bench format is:

```
entity TEST_BENCH is
end;

architecture TB_BEHAVIOR of TEST_BENCH is
  component ENTITY_UNDER_TEST
    port ( list-of-ports-their-types-and-modes );
  end component;
  Local-signal-declarations ;
begin
  Generate-waveforms-using-behavioral-constructs ;
  Apply-to-entity-under-test ;
  EUT : ENTITY_UNDER_TEST port map ( port-associations ) ;
  Monitor-values-and-compare-with-expected-values ;
end TB_BEHAVIOR;
```

Stimulus is automatically applied to the entity under test by instantiating the entity in the test bench model and then specifying the appropriate interface signals. The next two subsections look at waveform generation and output response monitoring.

11.2.1 Waveform Generation

There are two main approaches in generating stimulus values:

1. Create waveforms and apply stimulus at certain discrete time intervals.
2. Generate stimulus based on the state of the entity, that is, based on the output response of the entity.

Two types of waveforms are typically needed. One is a repetitive pattern, for example, in a clock, and the other is a sequential set of values.

Repetitive Patterns

A repetitive pattern with a constant on-off delay can be created using a concurrent signal assignment statement.

```
APR <= not APR after 20 ns;
                                    -- Signal APR is assumed to be of type BIT.
```

The waveform created is shown in Figure 11-1.

A clock with varying on-off period can be created using a process statement. The waveform created is that shown in Figure 11-1.

```
-- D_CLK is assumed to be a signal of type BIT.
process
  constant OFF_PERIOD: TIME := 30 ns;
  constant ON_PERIOD : TIME := 20 ns;
begin
  wait for OFF_PERIOD;
  D_CLK <= '1';
  wait for ON_PERIOD;
  D_CLK <= '0';
end process;
```

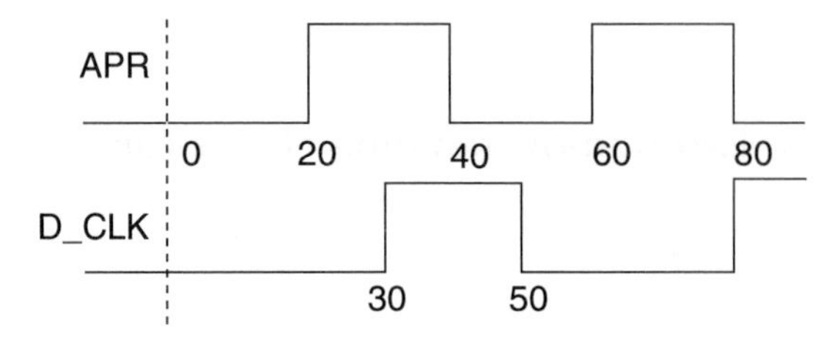

Figure 11-1 A repetitive pattern.

A problem with both of these approaches is that if the host environment provides no control over the amount of time simulation is to be performed, simulation would go on until time is equal to TIME'HIGH. One way to avoid this is to put an assertion statement in the process that will cause the assertion to fail after a specific time. Here is an example of such an assertion statement.

```
assert NOW <= 1000 ns
  report "Simulation completed successfully"
  severity ERROR;
```

Assertion would fail when simulation time exceeds 1000 ns, and simulation would stop (assuming that the simulator stops on getting a severity level of ERROR). Another way to stop a process from generating more events is by using a wait statement of the form:

```
wait;
```

The following `if` statement can be used in the above process to stop generating more events on D_CLK when simulation time exceeds 1000 ns.

```
if NOW > 1000 ns then
   wait;                         -- Suspend process indefinitely.
end if;
```

An alternate way to generate a clock with a varying on-off period is by using a conditional signal assignment statement. Here is an example.

```
D_CLK <=    '1' after OFF_PERIOD when D_CLK = '0' else
            '0' after ON_PERIOD;
-- D_CLK is a signal of type BIT.
-- OFF_PERIOD and ON_PERIOD are constants defined elsewhere.
```

A clock that is phase-delayed from another clock can be generated by using the 'DELAYED predefined attribute. For example,

```
DLY_D_CLK <= D_CLK'DELAYED (10 ns);
```

The waveforms for D_CLK and DLY_D_CLK are shown in Figure 11-2.

Two non-overlapping clocks can be generated in a similar manner by making sure that the time specified for the 'DELAYED attribute is larger than the clock's on period. For example,

```
NO_D_CLK <= D_CLK'DELAYED (25 ns);
```

will produce an non-overlapping clock. Waveforms are shown in Figure 11-2.

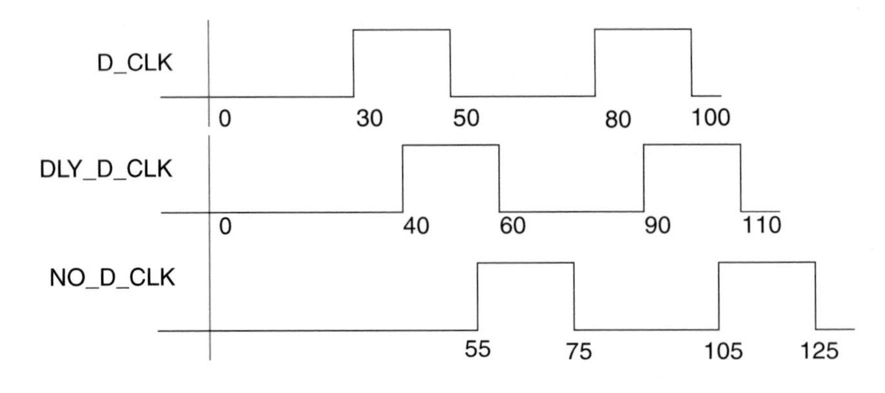

Figure 11-2 A phase-delayed clock and a non-overlapping clock.

A sequential set of values can also be generated for a signal by using multiple waveforms in a concurrent signal assignment statement. For example,

```
RESET <= '0', '1' after 100 ns, '0' after 180 ns, '1' after 210 ns;
```

The waveform generated on signal RESET is shown in Figure 11-3.

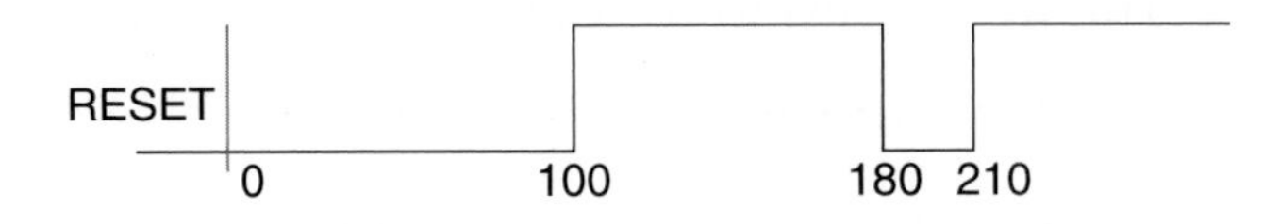

Figure 11-3 A non-repetitive waveform.

This waveform can be made to repeat itself by placing the signal assignment statement inside a process statement along with a `wait` statement. The following example of a two-phase clock shows such a repeated waveform. Figure 11-4 shows the waveforms created.

```
signal CLK1, CLK2: STD_LOGIC := '0';
. . .
TWO_PHASE: process
begin
   CLK1 <=     'U' after 5 ns, '1' after 10 ns, 'U' after 20 ns,
               '0' after 25 ns;
   CLK2 <=     'U' after 10 ns, '1' after 20 ns, 'U' after 25 ns,
               '0' after 30 ns;
   wait for 35 ns;
end process;
```

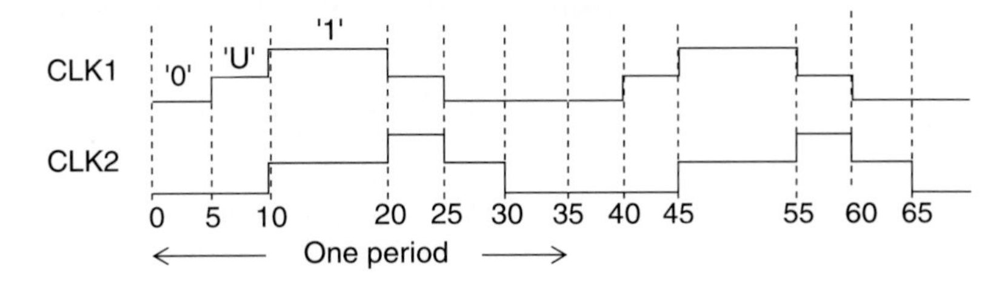

Figure 11-4 Complex repetitive waveforms.

Using Vectors

Another way to apply stimulus to a set of signals is to store the set of vectors in a constant table or in an ASCII file. Here is an example that stores the input vectors in a table.

```
constant NO_OF_BITS: INTEGER := 4;
constant NO_OF_VECTORS: INTEGER := 5;
type TABLE_TYPE is array (1 to NO_OF_VECTORS) of
                    STD_LOGIC_VECTOR(1 to NO_OF_BITS);
constant VECTOR_PERIOD: TIME := 100 ns;
constant INPUT_VECTORS: TABLE_TYPE :=
                    ("1001", "1000", "0010", "0000", "0110");
signal INPUTS: STD_LOGIC_VECTOR(1 to NO_OF_BITS);
signal A, B, C: STD_LOGIC;
signal D: STD_LOGIC_VECTOR(0 to 1);
```

Assume that the entity under test has four inputs, A, B, C, and D. If the vectors are to be applied at regular time intervals, a process statement can be used as shown.

```
process
  variable J: POSITIVE;                  -- Starts from 1
begin
  wait for VECTOR_PERIOD * J;
  INPUTS <= INPUT_VECTORS(J);
  J := J + 1;
  if J > NO_OF_VECTORS then
    wait;                                -- Suspend process indefinitely.
  end if;
end process;
A <= INPUTS(1);
B <= INPUTS(4);
C <= INPUTS(1);
D <= INPUTS(2 to 3);
```

If the vectors were to be applied at arbitrary intervals, a concurrent signal assignment statement with multiple waveforms can be used.

```
INPUTS <=     INPUT_VECTORS(1) after 10 ns,
              INPUT_VECTORS(2) after 25 ns,
              INPUT_VECTORS(3) after 30 ns,
              INPUT_VECTORS(4) after 32 ns,
              INPUT_VECTORS(5) after 40 ns;
```

The input vectors could also be specified in an ASCII file as a sequence of values. The following statements define a file and read the vectors from the file.

```
process
  type VEC_TYPE is file of STD_LOGIC_VECTOR;
  file VEC_FILE: VEC_TYPE open READ_MODE
    is "/usr/jb/EXAMPLE.vec";
  variable LENGTH: INTEGER;
  variable IN_VECTOR: STD_LOGIC_VECTOR(1 to 4);
begin
  while not ENDFILE(VEC_FILE) loop
    READ (VEC_FILE, IN_VECTOR, LENGTH);
    -- The READ operation returns the vector read in IN_VECTOR
    -- and the number of elements in the vector in LENGTH.
    :
    :
  end loop;
end process;
```

A complete test bench that uses the waveform application method is shown next. The entity being simulated is called DIV. The output values are written into a file for later comparison.

```
library IEEE;
use IEEE.STD_LOGIC_1164.all;
entity DIV_TB is end;

architecture DIV_TB_BEH of DIV_TB is
  component DIV
    port ( CK, RESET, TESTN: in STD_LOGIC;
           ENA: out STD_LOGIC);
  end component;
  signal CLOCK, RESET, TESTN, ENABLE: STD_LOGIC;
  type VEC_TYPE is file of STD_LOGIC;
  file OUTFILE: VEC_TYPE open WRITE_MODE
    is "/usr1/jb/div.vec.out";
  for D1: DIV use entity WORK.DIV;
begin
```

```
    CKP: process
    begin
      CLOCK <= '0';
      wait for 3 ns;
      CLOCK <= '1';
      wait for 5 ns;
      if NOW > 900 ns then
        report "Simulation completed successfully.";
        wait;                                       -- Stop simulation.
      end if;
    end process CKP;

    RESET <= '0', '1' after 110 ns;
    TESTN <= '0', '1' after 160 ns, '0' after 670 ns;
    -- Apply to entity under test:
    D1: DIV port map (CLOCK, RESET, TESTN, ENABLE);

    -- For every event on the ENABLE output signal, write to file.
    MONITOR: process (ENABLE)
    begin
      WRITE (OUTFILE, ENABLE);
    end process MONITOR;
  end DIV_TB_BEH;
```

In the second approach for stimulus generation, the stimulus value generated is based on the state of the entity under test. This approach is useful in testing a finite state machine for which different input stimulus is applied based on the machine's state. Consider an entity in which the objective is to compute the factorial of an input number. The handshake mechanism between the entity under test and the test bench model is shown in Figure 11-5.

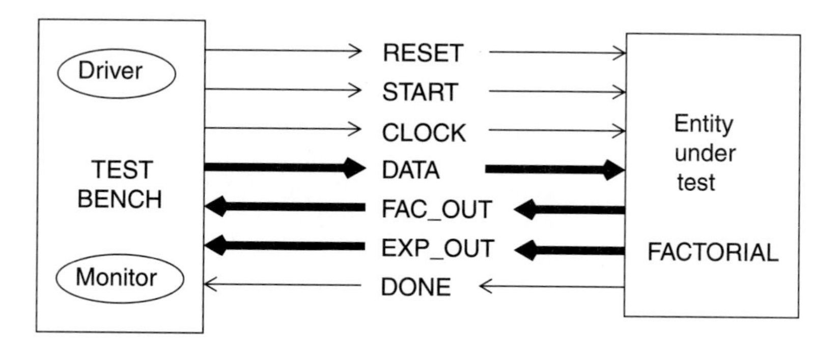

Figure 11-5 Handshake between test bench and entity under test.

The RESET input to the entity resets the factorial model to an initial state. The START signal is set after the DATA input is applied. When computation is complete, the output DONE signal is set to indicate that the computed result appears on the FAC_OUT and EXP_OUT outputs. The resulting factorial value is FAC_OUT * 2^{EXP_OUT}. The test bench model provides input data on signal DATA starting from values 1 to 20 in increments of one. It applies the data, sets the START signal, waits for the DONE signal, and then applies the next input data. Assertion statements are used to make sure that the values appearing at the output are correct. The test bench description follows.

```
library IEEE, BOOK_LIB;
use IEEE.STD_LOGIC_1164.all, BOOK_LIB.UTILS_PKG.all;
  -- Functions TO_INTEGER and TO_STDLOGICVECTOR used in
  -- model are defined in package UTILS_PKG that resides in design
  -- library BOOK_LIB.
entity FAC_TB is
  constant IN_MAX: INTEGER := 5;
  constant OUT_MAX: INTEGER := 8;
end FAC_TB;

architecture FAC_TB_FUNC of FAC_TB is
  component FACTORIAL
    port (
      RESET, START, CLOCK: in STD_LOGIC;
      DATA: in STD_LOGIC_VECTOR(IN_MAX–1 downto 0);
      DONE: out STD_LOGIC;
      FOUT,
      EOUT: out STD_LOGIC_VECTOR(OUT_MAX–1 downto 0));
  end component;
  type FAC_STATE is (RESET_ST, START_ST,
            APPL_DATA_ST, WAIT_RESULT_ST);
  signal CLK, RESET, START, DONE: STD_LOGIC;
  signal DATA:
    STD_LOGIC_VECTOR(IN_MAX–1 downto 0) := (others => '0');
  signal FAC_OUT, EXP_OUT:
            STD_LOGIC_VECTOR(OUT_MAX–1 downto 0);
  signal NEXT_STATE: FAC_STATE;          -- Starts with RESET_ST state.
  constant MAX_APPLY: POSITIVE := 20;
begin
  CLK_P: process
  begin
    CLK <= '0';
    wait for 4 ns;
    CLK <= '1';
```

```
    wait for 6 ns;
  end process CLK_P;

  process
    variable NUM_APPLIED: POSITIVE;                -- Starting with 1.
  begin
    if CLK = '0' and CLK'EVENT then              -- Falling edge transition.
      case NEXT_STATE is
        when RESET_ST =>
          RESET <= '1'; START <= '0';
          NEXT_STATE <= APPL_DATA_ST;
        when APPL_DATA_ST =>
          DATA <= TO_STDLOGICVECTOR
                    (NUM_APPLIED, IN_MAX);
          NEXT_STATE <= START_ST;
        when START_ST =>
          START <= '1';
          NEXT_STATE <= WAIT_RESULT_ST;
        when WAIT_RESULT_ST =>
          RESET <= '0'; START <= '0';
          wait until DONE = '1';
          assert (NUM_APPLIED = TO_INTEGER(FAC_OUT) *
                (2 ** TO_INTEGER(EXP_OUT)))
            report "Incorrect result from factorial model.";
          NUM_APPLIED := NUM_APPLIED + 1;
          if NUM_APPLIED < MAX_APPLY then
            NEXT_STATE <= APPL_DATA_ST;
          else
            assert FALSE                         -- Stop simulation.
              report "Test completed successfully.";
          end if;
      end case;
    end if;
  end process;

  -- Apply to entity under test:
  F1: FACTORIAL port map (RESET, START, CLK, DATA,
                          DONE, FAC_OUT, EXP_OUT);
end FAC_TB_FUNC;
```

11.2.2 Monitoring Behavior

In the test bench model for the FACTORIAL entity, we saw how the test bench can monitor the behavior of the entity and apply different patterns based on the output response. Another common way to monitor the output response of an entity is to apply a vector, sample the output after a specific time, and then verify to make sure that the output response matches the expected values. Here is an example of such a test bench. The input and output vectors are stored in tables. Alternately, they could have been read from an ASCII file.

```
library IEEE;
library BOOK_LIB;
use IEEE.STD_LOGIC_1164.all;
use BOOK_LIB.UTILS_PKG.all;
entity ANOTHER_DIV_TB is end;

architecture APPLY_AND_SAMPLE of ANOTHER_DIV_TB is
  component DIV
    port (  CK, RESET, TESTN: in STD_LOGIC;
            ENA: out STD_LOGIC);
  end component;
  type SLV3 is array (1 to 3) of STD_LOGIC;
  type SLV2_VECTOR is array (POSITIVE range <>,
    POSITIVE range <>) of STD_LOGIC;
  constant INPUT_VECTORS: SLV2_VECTOR :=
    ("100", "100", "100", "100", "110", "111", "011");
  constant OUTPUT_VECTORS: SLV2_VECTOR :=
    ("0", "0", "0", "0", "1");
  constant STROBE_DELAY: TIME := 1 ns;
  constant CYCLE_TIME: TIME := 80 ns;
  signal CLOCK, RESET, TESTN, ENABLE: STD_LOGIC;
begin
  APPLY: process
  begin
    for J in INPUT_VECTORS'RANGE(1) loop
      GENERATE_PULSE (
        DRIVE_SIGNAL => CLOCK,
        SIGNAL_VALUE => INPUT_VECTORS(J, 1),
        DRIVE_DELAY => 20 ns,
        DRIVE_WIDTH => 30 ns);
      -- Procedure GENERATE_PULSE generates a pulse of specified
      -- delay and width; declared in package UTILS_PKG.
      RESET <= INPUT_VECTORS(J, 2);
      TESTN <= INPUT_VECTORS(J, 3);
      wait for CYCLE_TIME – STROBE_DELAY;
      assert ENABLE = OUTPUT_VECTORS(J, 1);
      wait for STROBE_DELAY;
    end loop;
  end process APPLY;
```

```
    -- Entity under test:
    D2: DIV port map (CLOCK, RESET, TESTN, ENABLE);
end APPLY_AND_SAMPLE;
```

In this test bench example, the application of test vectors is expressed as a process. After a test vector is applied, the process suspends for time (CYCLE_TIME – STROBE_DELAY), samples the output, and then checks to see if the expected output is equal to the specified output vector. The process then suspends for STROBE_DELAY time before reapplying the next vector.

11.3 Converting Real and Integer to Time

If an integer needs to be converted to a time value, one way is to multiply the integer with the required time unit. Here is an example.

```
variable INCR: INTEGER;
. . .
EFF_A <= DRV_A after INCR * 1 ns;          -- Delay is in ns.
wait for (INCR / 2) * 1 ms;                -- Delay is in ms.
```

Variable INCR is of type INTEGER. By multiplying it by 1 ns, the expression "INCR * 1 ns" becomes a value of type TIME. Similarly, the integer expression "INCR / 2" has been changed to a time value by multiplying it by 1 ms.

A time value can be converted to an integer by simply doing the opposite, that is, dividing the time value by a time unit.

```
17 ns / 1 ns                               -- Gives integer 17
```

Similarly, a real value can be converted to time by multiplying the real value by a time unit.

```
52.44 * 1 ns                               -- A time value
```

11.4 Dumping Results into a Text File

It is possible to write out the values of signals and variables at specific times into a text file by using the WRITE procedures provided in the predefined TEXTIO package. A simulator usually provides this capability by using a command file that contains a list of signals whose values need to be saved; this capability, however, is external to the VHDL model.

Here is a skeleton of an architecture body that prints out a dump file which contains a header and a set of values on every clock transition.

```
library IEEE, BOOK_LIB;
use STD.TEXTIO.all;            -- This context clause has to be included
          -- because types and procedures from package TEXTIO are used.
use IEEE.STD_LOGIC_1164.all, BOOK_LIB.UTILS_PKG.all;
  -- Package UTILS_PKG contains the TO_STRING function.
architecture TEST of TOP is
  -- First define a file DUMP:
  file DUMP: TEXT open WRITE_MODE is "/home/jb/top.dump";
  -- TEXT is a file type predefined in the TEXTIO package.
  component DESIGN_BEING_MONITORED port (
      InputA: in BIT_VECTOR(0 to 7);
      InputB: in INTEGER;
      InputC: in BIT;
      OutputA: out STD_LOGIC_VECTOR(3 downto 0);
      OutputB: out REAL);
  end component;
  signal N1: BIT_VECTOR(0 to 7);
  signal N2: INTEGER;
  signal CLK: BIT;
  signal T1: STD_LOGIC_VECTOR(3 downto 0);
  signal T2: REAL;
begin
  -- Design to be monitored:
  D1: DESIGN_BEING_MONITORED port map (
      InputA => N1,
      InputB => N2,
      InputC => CLK,
      OutputA => T1,
      OutputB => T2);

  process
    variable BUF: LINE;            -- LINE is an access type predefined
                                   -- in the TEXTIO package.
  begin
    -- Write header only in the beginning:
    WRITE (BUF,
        STRING'("Following is a dump of the circuit TOP:"));
    WRITELINE (DUMP, BUF);
```

```
    loop
      wait on CLK;            -- Loop will execute on every edge of clock.
      WRITE (BUF, STRING'("TIME = "));          -- Write STRING.
      WRITE (BUF, NOW);                -- Write current simulation time.
      WRITE (BUF, STRING'(", CLK = "));
      WRITE (BUF, CLK);                         -- Write BIT.
      WRITE (BUF, STRING'(", N1 = "));
      WRITE (BUF, N1);                          -- Write BIT_VECTOR.
      WRITE (BUF, STRING'(", N2 = "));
      WRITE (BUF, N2);                          -- Write integer.
      WRITE (BUF, STRING'(", T1 = "));
      WRITE (BUF, TO_STRING (T1));
        -- Writes STD_LOGIC_VECTOR value T1.
        -- Function TO_STRING converts a STD_LOGIC_VECTOR
        -- to a STRING.
      WRITE (BUF, STRING'(", T2 = "));
      WRITE (BUF, T2);                          -- Write real.
      WRITELINE (DUMP, BUF);                    -- Write line to file.
    end loop;
  end process;
end TEST;
```

The output looks like this.

```
Following is a dump of the circuit TOP:
Time = 10 ns, CLK = 0, N1 = 01101110, N2 = 7, T1 = 0ZZZ, T2 =1.056000e+02
Time = 12 ns, CLK = 1, N1 = 01101111, N2 = 2, T1 = 0Z00, T2 =2.300000e-01
Time = 14 ns, CLK = 0, N1 = 11101110, N2 = 8, T1 = 0ZZZ, T2 =1.000000e+00
. . . and so on.
```

A context clause using the TEXTIO package is specified with the architecture body. A file DUMP is declared of the predefined file type TEXT. A variable BUF is defined of the predefined access type LINE. Values are written to BUF using the predefined WRITE procedures. After all the characters to appear on a line are written to BUF, its contents are written out to the file DUMP using the predefined WRITELINE procedure. The WRITELINE procedure also clears out the BUF variable so that the next WRITE procedure will start writing into BUF from the beginning of a line. The procedure WRITE is predefined in the TEXTIO package for the predefined types BIT, BOOLEAN, CHARACTER, STRING, BIT_VECTOR, INTEGER, REAL, and TIME. Therefore, for each different predefined type, a different WRITE procedure is used. For example,

```
WRITE (BUF, CLK);
```

is the WRITE procedure that writes a BIT value. Another example is:

```
WRITE (BUF, STRING'(", CLK = "));
```

which is a call to the WRITE procedure that writes a STRING value. Note that a conversion function is needed to print a value of a type other than the predefined types. This is done in the architecture body for STD_LOGIC_VECTOR values by using the conversion function TO_STRING, which converts a STD_LOGIC_VECTOR value to a STRING value.

In the above architecture body, real numbers are printed in the exponent notation form. If real numbers need to be printed in decimal form, a value for the default parameter DIGITS in the WRITE procedure has to be specified. This parameter value specifies the number of digits to be used to the right of the decimal point. By changing the WRITE procedure call (that writes value of T2) to:

```
WRITE (L => BUF, VALUE => T2, DIGITS => 4);
```

we get the output:

```
Following is a dump of the circuit TOP:
Time = 10 ns, CLK = 0, N1 = 01101110, N2 = 7, T1 = 0ZZZ, T2 =105.6000
Time = 12 ns, CLK = 1, N1 = 01101111, N2 = 2, T1 = 0Z00, T2 =0.2300
Time = 14 ns, CLK = 0, N1 = 11101110, N2 = 8, T1 = 0ZZZ, T2 =1.0000
. . . and so on.
```

There are many other ways of making this dump process more modular. A procedure could be defined that dumps the values of its parameters into a specified file, such as:

```
procedure DUMP_PROC (file D_FILE: TEXT; CLOCK: in BIT;
   EN1: in BIT_VECTOR; EN2: in INTEGER;
   TE1, TE2: in STD_LOGIC_VECTOR);
```

This procedure would contain the entire set of statements that appear in the process. The process statement would then contain just the procedure call:

```
DUMP_PROC (DUMP, CLK, N1, N2, T1, T2);
```

Conversion functions can be avoided by writing overloaded procedures for WRITE that operate on STD_LOGIC_VECTOR types, such as:

```
procedure WRITE (   L: inout LINE;
                    VALUE: in STD_LOGIC_VECTOR);
```

If the dump needs to be produced, say, on every falling clock edge, the wait statement in the process can be modified to:

```
wait until CLK = '0';
```

A dump can also be produced directly at the standard output of a terminal by using the predefined file OUTPUT. If the WRITELINE procedure call in the previous process were changed to:

```
WRITELINE (OUTPUT, BUF);          -- Write line to standard output.
```

all output would appear at the terminal (standard output).

Similarly, standard input from the keyboard can be received by using the predefined file INPUT.

For a line that is being written to, such as BUF in the previous process statement, BUF'LENGTH gives the number of characters that have been written to the line. This feature can be used, for example, to start a new line if more than 80 characters have been printed on one line.

```
-- If more than 80 characters loaded in buffer:
if BUF'LENGTH > 80 then
   -- Dump the current line.
   WRITELINE (OUTPUT, BUF);
end if;
-- Start a new line.
```

Converting Integer to String

In the previous chapter, we saw how the 'IMAGE attribute is used to convert integers (and values of other types) to strings. The WRITE procedure defined in the TEXTIO package can be used to convert integers (also real, BIT and BOOLEAN) into strings. Here is an example.

```
variable LINE_BUF: LINE;
variable STR_BUF: STRING(1 to 20) := (others => ' ');
variable INT_VAL: INTEGER;          -- The integer value to be converted.
. . .
-- Write the integer into the line buffer:
WRITE (LINE_BUF, INT_VAL);
   -- Automatically allocates sufficient memory to store INT_VAL.

-- Copy into string buffer:
STR_BUF(1 to LINE_BUF'LENGTH) := LINE_BUF.all;
```

```
-- If LINE_BUF is not required any more, then need to
-- deallocate memory that it is pointing to:
DEALLOCATE (LINE_BUF);
```

The WRITE procedure writes the integer value into the line buffer. Sufficient memory is automatically allocated by the WRITE procedure. The assignment statement causes the string pointed by LINE_BUF to be copied into the string STR_BUF. The DEALLOCATE procedure deallocates the memory that was allocated by the WRITE procedure (a DEALLOCATE procedure is implicitly defined for every access type).

11.5 Reading Vectors from a Text File

It is possible to read stimulus from a text file using READ procedures provided in the predefined TEXTIO package. Here is an example that reads the stimulus from a file, one line at a time, and applies the appropriate values from the line to the inputs of the entity under test.

```
library IEEE, BOOK_LIB;
use IEEE.STD_LOGIC_1164.all;
use STD.TEXTIO.all;                       -- Include TEXTIO package.
use BOOK_LIB.UTILS_PKG.all;
  -- Function TO_STDLOGICVECTOR is defined in
  -- package UTILS_PKG that resides in library BOOK_LIB.
architecture TEST of BOTTOM is
  -- Specify vector file:
  file VECTORS: TEXT open READ_MODE is "alu3.vec";
  -- This file contains lines of the form (without the two dashes):
  --   # This is a comment line.
  --   0001 81 11100100
  --   0100 56 10000011
  --   . . .
  -- Declare the entity under test:
  component EUT
    port ( EN1: in BIT_VECTOR (0 to 3); EN2: in INTEGER;
           EPX: in STD_LOGIC_VECTOR (7 downto 0); . . . );
  end component;
  constant VECTOR_APPLICATION_PERIOD : TIME := 5 ns;
  signal N1: BIT_VECTOR (0 to 3);
  signal N2: INTEGER;
  signal PARX: STD_LOGIC_VECTOR (7 downto 0);
```

```
begin
  process
    variable BUF: LINE;
    variable N1_VAR: BIT_VECTOR (0 to 3);
    variable N2_VAR: INTEGER;
    variable PARX_STR: STRING (8 downto 1);
  begin
    while not ENDFILE (VECTORS) loop
      READLINE (VECTORS, BUF);          -- Read line from file.
      if BUF(1) = '#' then
        next;                           -- Skip this line in file.
      end if;
      READ (BUF, N1_VAR);               -- Read BIT_VECTOR.
      READ (BUF, N2_VAR);               -- Read integer.
      READ (BUF, PARX_STR);             -- Read STRING.
      -- Apply values read to input signals of entity under test:
      N1 <= N1_VAR;
      N2 <= N2_VAR;
      PARX <= TO_STDLOGICVECTOR (PARX_STR);
      wait for VECTOR_APPLICATION_PERIOD;
    end loop;
    wait;                               -- Stop simulation.
  end process;

  -- Instantiate the entity under test:
  E1: EUT port map (EN1 => N1, EN2 => N2, EPX => PARX, . . .);
  . . .
end TEST;
```

File VECTORS is declared to be of the predefined file type TEXT. The READLINE procedure reads one line from the file VECTORS into the variable BUF. The values in BUF are read using the READ procedures. The READ procedures are present in the predefined TEXTIO package for the predefined types BIT, CHARACTER, BOOLEAN, INTEGER, REAL, BIT_VECTOR, TIME, and STRING. The first READ procedure reads a 4-bit BIT_VECTOR value into the variable N1_VAR. A temporary variable is necessary because a signal cannot be passed in as a parameter to the READ procedure. The variable is later assigned to the input signal N1 of the entity under test.

The second READ procedure reads an INTEGER value from BUF. The third READ procedure reads an 8-bit STRING value from BUF. This string value is converted to a STD_LOGIC_VECTOR value using a function. This is necessary because a READ procedure is not predefined to operate on STD_LOGIC_VECTOR values. Alternately, an overloaded READ procedure could have been defined to read STD_LOGIC_VECTOR values directly, such as:

```
procedure READ (L: inout LINE;
                        VALUE: out STD_LOGIC_VECTOR);
```

In such a case, no function is necessary to convert from STRING to STD_LOGIC_VECTOR values.

After the values from one line in the input file are applied to the inputs of the entity under test, the next iteration of the `while` loop starts after the vector application period. If no more vectors are present in the input file, the function ENDFILE returns true and the `while` loop terminates.

Standard input from the keyboard can be received by using the predefined file INPUT. For example, replacing the READLINE procedure call in the previous process by:

```
READLINE (INPUT, BUF);          -- Read line from keyboard.
```

causes all input to be read from a keyboard (standard input).

For a line that is being read, such as BUF in the previous process statement, BUF'LENGTH gives the number of characters still to be read. BUF'LENGTH is 0 when end of line is reached. Here is an example that reads all integers on a line until end-of-line is reached.

```
variable BUF: LINE;
file IN_FILE: TEXT open READ_MODE is "strings.vec";
variable INT_VAR: INTEGER;
. . .
READLINE (IN_FILE, BUF);
while (BUF /= NULL) and (BUF'LENGTH /= 0) loop
  READ (BUF, INT_VAR);
  . . .
end loop;
```

Every time procedure READ is called, the access pointer BUF moves past the integer just read; consequently BUF'LENGTH decreases and contains the number of characters still remaining on the line to be read.

11.6 A Test Bench Example

Here is a test bench for a full-adder circuit. The input vectors are read from a text file and results are dumped into a text file.

```
library BOOK_LIB;
use BOOK_LIB.UTILS_PKG.all;
use STD.TEXTIO.all;
entity FA_TEST is end;

architecture IO_EXAMPLE of FA_TEST is
  component FULL_ADD
    port (CIN, A, B: in BIT; COUT, SUM: out BIT);
  end component;

  file VEC_FILE: TEXT
    open READ_MODE is "/home/bhasker/MODELS/fadd.vec";
  file RESULT_FILE: TEXT
    open WRITE_MODE is "/home/bhasker/MODELS/fadd.out";
  signal S: BIT_VECTOR(0 to 2);
  signal Q: BIT_VECTOR(0 to 1);
begin
  FA: FULL_ADD port map (S(0), S(1), S(2), Q(0), Q(1));

  process
    constant PROPAGATION_DELAY: TIME := 2 ns;
    variable BUF_IN, BUF_OUT: LINE;
    variable OUT_STR: BIT_VECTOR(0 to 1);
    variable NUM_VECTORS: INTEGER := 0;
    variable VAR_S: BIT_VECTOR(0 to 2);
  begin
    while not ENDFILE (VEC_FILE) loop
      READLINE (VEC_FILE, BUF_IN);
      -- Make sure BUF contains only 3 characters:
      assert (BUF_IN'LENGTH = 3)
        report "Vector does not have three bits"
        severity ERROR;
      READ(BUF_IN, VAR_S);          -- Read BIT_VECTOR from line.
      S <= VAR_S;
      wait for PROPAGATION_DELAY;
      NUM_VECTORS := NUM_VECTORS + 1;
      WRITE (BUF_OUT, NUM_VECTORS);
      WRITE (BUF_OUT, STRING'(". "));
      WRITE (BUF_OUT, S);
      WRITE (BUF_OUT, STRING'(" ----> "));
```

```
        OUT_STR := Q;
        WRITE (BUF_OUT, OUT_STR);
        WRITELINE (RESULT_FILE, BUF_OUT);
      end loop;
      report "Completed processing all vectors.";
      wait;                                          -- Stop simulation.
    end process;
  end IO_EXAMPLE;
```

Note the temporary assignment of Q to OUT_STR. This is necessary since only a variable can be passed into a WRITE procedure. Similarly a temporary variable VAR_S is required for S since only a variable can be passed to a READ procedure. Here is the input for the test bench.

```
000
001
010
011
100
101
110
111
```

The following is the output produced by this test bench.

```
1. 000 ----> 00
2. 001 ----> 01
3. 010 ----> 01
4. 011 ----> 10
5. 100 ----> 01
6. 101 ----> 10
7. 110 ----> 10
8. 111 ----> 11
```

An example of a test bench that reads from standard input and writes to standard output is described in Section 12.15. A test bench is also described in Section 12.16 for a pulse counter example.

11.7 Initializing a Memory

Assume there exists an N-by-M memory and we need to initialize or load it with values. A memory can be represented using either a signal or a variable; here we use a signal.

```
constant N: POSTIVE := 256;          -- Number of words in memory.
constant M: POSITIVE := 8;           -- Number of bits per word.
type MTYPE is array (0 to N–1) of STD_LOGIC_VECTOR(0 to M–1);

signal MEMORY: MTYPE;                -- The memory.
```

Initializing the same value to all bits of memory can be achieved by simply assigning an aggregate. Here is an example that assigns '0' to all bits.

```
signal MEMORY : MTYPE := (others => (others =>'0'));
```

What if the values are to be loaded from a text file? The load file has entries of the form:

```
00010000
00111001
11010100
11110001
. . .
```

The first word is to be loaded at address 0, second word in address 1, and so on. Here is a process that reads such a load file.

```
library IEEE, BOOK_LIB;
use STD.TEXTIO.all;
use IEEE.STD_LOGIC_1164.all, BOOK_LIB.UTILS_PKG.all;
architecture LOAD of LOAD_MEMORY is
  constant N: POSITIVE := 256;         -- Number of words in memory.
  constant M: POSITIVE := 8;           -- Number of bits per word.
  type MTYPE is array (0 to N–1) of
        STD_LOGIC_VECTOR(0 to M–1);
  signal MEMORY: MTYPE;                -- The memory.
begin
  process
    variable INBUF: LINE;
    file LOAD_FILE: TEXT open READ_MODE
        is "/home/bhasker/ram1.load";
    variable NUM_WORDS : INTEGER := 0;
    variable WORD: STRING(1 to M);
  begin
```

```
    while not ENDFILE (LOAD_FILE) loop
      READLINE (LOAD_FILE, INBUF);
      READ (INBUF, WORD);                 -- Read the word as a string.
      MEMORY (NUM_WORDS) <=
          TO_STDLOGICVECTOR(WORD);
        -- TO_STDLOGICVECTOR is defined in package
        -- UTILS_PKG. It converts a string value to a
        -- STD_LOGIC_VECTOR type.
      NUM_WORDS := NUM_WORDS + 1;
      if NUM_WORDS > N then
        exit;                 -- All addresses in memory loaded; quit loop.
      end if;
    end loop;
    wait;
  end process;
end LOAD;
```

What if we need to load only words at specific locations in memory? The load file in this case may look like this:

```
8 00101010
15 11100101
22 11110000
...
```

The first entry is the address and the vector following it is the value to be loaded into that location. The addresses need not be in any order. Here is the modified process statement to read this load file.

```
PARTIAL_LOAD: process
  variable INBUF: LINE;
  file LOAD_FILE: TEXT open READ_MODE
      is "/home/bhasker/ram1.partial_load";
  variable ADDRESS : NATURAL;
  variable WORD: STRING(1 to M);
begin
  while not ENDFILE (LOAD_FILE) loop
    READLINE (LOAD_FILE, INBUF);
    READ (INBUF, ADDRESS);            -- Read the address first.
    READ (INBUF, WORD);               -- Read the word.
    MEMORY (ADDRESS) <=
          TO_STDLOGICVECTOR(WORD);
  end loop;
  wait;
end process;
```

11.8 Variable File Names

Instead of hard-coding a file name in a VHDL model, it is possible to use a variable file name and provide different file names in multiple simulation runs.

11.8.1 File Name as a Generic

One approach is to use a generic for a file name, as shown in the following example.

```
use STD.TEXTIO.all;
entity GENERIC_FILE_TB is
  generic (FILE_NAME: STRING);
end GENERIC_FILE_TB;

architecture TB of GENERIC_FILE_TB is
begin
  process
    variable LINE_BUF: LINE;
    variable FILE_STATUS: FILE_OPEN_STATUS;
    file FILE_PTR: TEXT;
  begin
    -- Example of opening a file for reading:
    FILE_OPEN (
      F => FILE_PTR,
      EXTERNAL_NAME => FILE_NAME,
      OPEN_KIND => READ_MODE,
      STATUS => FILE_STATUS);

    assert FILE_STATUS = OPEN_OK
      report "Something is wrong in opening file " & FILE_NAME &
             " in read mode."
      severity ERROR;

    -- As an example, read line from file and write to standard output:
    READLINE (FILE_PTR, LINE_BUF);
    WRITELINE (OUTPUT, LINE_BUF);
    wait;
  end process;
end TB;
```

When this test bench is instantiated or during compilation (some compilers provide the capability of specifying a value for a generic at run-time), the file name can be supplied. A default value for the generic could also be provided in the generic declaration such as:

```
generic (FILE_NAME: STRING := "/home/bhasker/mem.vec");
```

11.8.2 File Name from Standard Input

Another approach is to read the file name from standard input. Here is an example.

```
STDIN: process
  variable FILEBUF, LINEBUF: LINE;
  file CMDFILE: TEXT;
  variable FILE_STATUS: FILE_OPEN_STATUS;
begin
  WRITE (LINEBUF, STRING'("Enter file name:"));
  WRITELINE (OUTPUT, LINEBUF);          -- The string will appear on
                                        -- standard output.
  READLINE (INPUT, FILEBUF);            -- File name is read from
                                        -- standard input.
  -- Open file for writing:
  FILE_OPEN (F => CMDFILE,
    EXTERNAL_NAME => FILEBUF.all,
    OPEN_KIND => WRITE_MODE,
    STATUS => FILE_STATUS);

  assert FILE_STATUS = OPEN_OK
    report "Cannot open " & FILEBUF.all & " for writing."
    severity ERROR;

  -- File CMDFILE is ready for write operation.
  -- As an example, contents of FILEBUF is written to CMDFILE:
  WRITELINE (CMDFILE, FILEBUF);
  wait;
end process;
```

The first two statements in the process cause the string "Enter file name:" to be printed to standard output. The READLINE procedure reads the string entered at standard input. The file is opened for writing; FILEBUF.**all** refers to the entire contents of the FILEBUF (note that FILEBUF is a variable of an access type) which should be the file name entered at standard input. The last procedure call writes out the contents of FILEBUF (the file name) to file CMDFILE.

❑

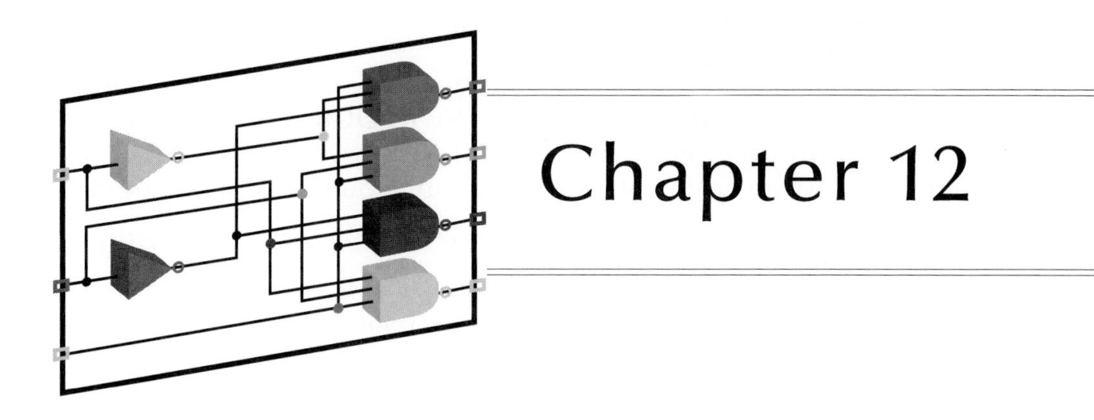

Chapter 12

Hardware Modeling Examples

This chapter describes a number of examples of hardware models using VHDL. As we have seen, there is more than one way to model a particular entity using this language. In this chapter we present one such approach for each entity being modeled. We shall be using the type STD_LOGIC defined in package STD_LOGIC_1164 for all the hardware models that are described in this chapter. The complete description of the STD_LOGIC_1164 package appears in Appendix E.

12.1 Modeling Entity Interfaces

The interface ports of an entity are specified using the entity declaration. The ports specify the names of signals that interact with the external environment of the entity. The semantics of each interface port is derived from the mode and type of the port. Here is an example of the external interface for an or-and-invert entity.

```
library IEEE; use IEEE.STD_LOGIC_1164.all;
entity OAI32 is
```

```
   port (A1, A2, A3, B1, B2: in STD_LOGIC; Z: out STD_LOGIC);
end OAI32;
```

This entity declaration specifies that the entity OAI32 has five input ports and one output port, all of type STD_LOGIC, that is, the ports can have any of the nine values 'U', 'X', '0', '1', 'Z', 'W', 'L', 'H', or '-'. An example of an entity interface for a 4-bit counter is shown next.

```
library IEEE; use IEEE.STD_LOGIC_1164.all;
entity COUNTER4 is
   port (COUNT, CLOCK: in STD_LOGIC;
         Q: out STD_LOGIC_VECTOR(3 downto 0));
end COUNTER4;
```

This entity declaration declares two 1-bit inputs and a 4-bit output for entity COUNTER4. Its external view is shown in Figure 12-1.

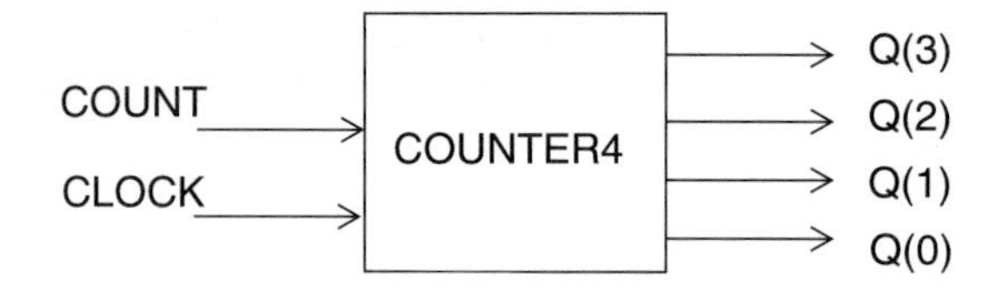

Figure 12-1 An external view of a 4-bit counter.

12.2 Modeling Simple Elements

A basic hardware element is a wire. A wire can be modeled in VHDL as a signal. Consider a 4-bit and gate the behavior of which is described next.

```
library IEEE; use IEEE.STD_LOGIC_1164.all;
entity AND4 is end;

architecture AND4_DF of AND4 is
   signal A, B, C: STD_LOGIC_VECTOR(3 downto 0);
begin
   A <= B and C after 5 ns;
end AND4_DF;
```

The gate delay for the and gate is specified to be 5 ns. An interesting point to note is that the previous architecture body is a legal VHDL description inspite of the fact that the entity has no inputs or outputs. The hardware represented by this model is shown in Figure 12-2.

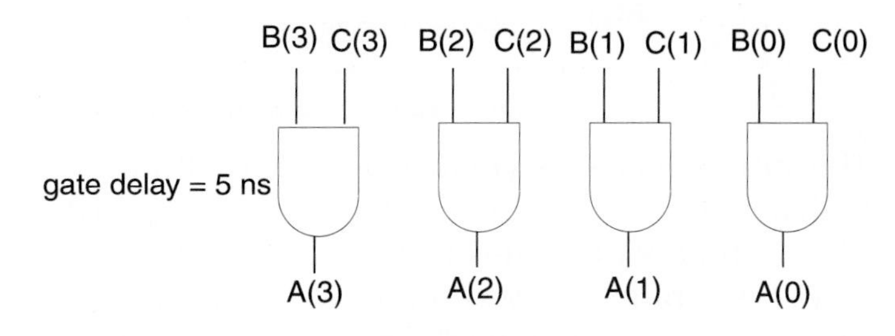

Figure 12-2 A 4-bit and gate.

This example and the one following show that Boolean equations can be modeled as expressions in concurrent signal assignment statements. Wires can be modeled as signal objects. For example, in the following description signal F represents a wire that connects the output of the **not** operator to the input of the **xor** operator. The circuit represented by the architecture is shown in Figure 12-3.

```
architecture BOOLEAN_EX of B_EX is
  signal D, E, F, G: STD_LOGIC;
begin
  D <= F xor G;
  F <= not E;
end BOOLEAN_EX;
```

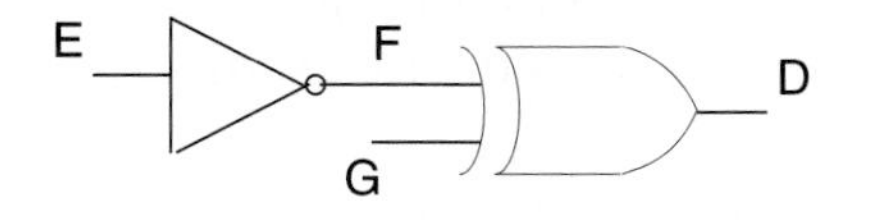

Figure 12-3 A combinational circuit.

Consider the following behavior and its corresponding hardware representation, as shown in Figure 12-4.

```
architecture LOOP1 of ASYNCHRONOUS is
  signal APQ, BQT, CTP, DZE: STD_LOGIC;
begin
  CTP <= APQ or DZE;
  APQ <= BQT nand CTP;
end LOOP1;
```

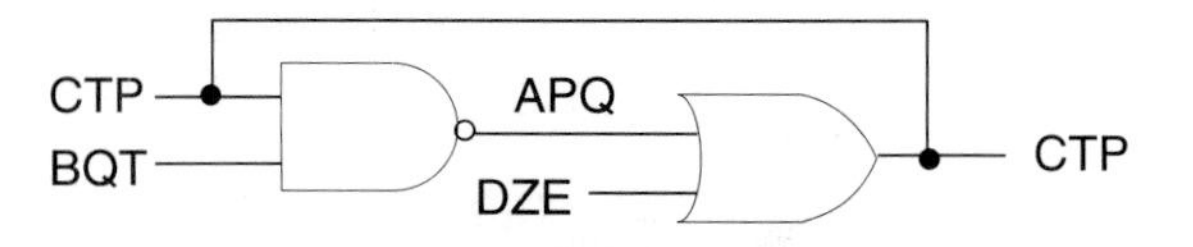

Figure 12-4 An asynchronous loop.

This circuit has an asynchronous loop. If the model were simulated with a certain set of signal values (BQT = '1', CTP = '1', DZE = '0'), simulation time would never advance because the simulator would always be iterating between the two signal assignments. The iteration time would be two delta delays. Therefore, extra caution must be exercised when values are assigned to signals and when these same signals are used in expressions.

In certain cases, it is desirable to have such an asynchronous loop. An example of such an asynchronous loop is shown next; the statement represents a periodic waveform with a cycle of 20 ns. Its hardware representation is shown in Figure 12-5.

```
ACQ <= not ACQ after 10 ns;
```

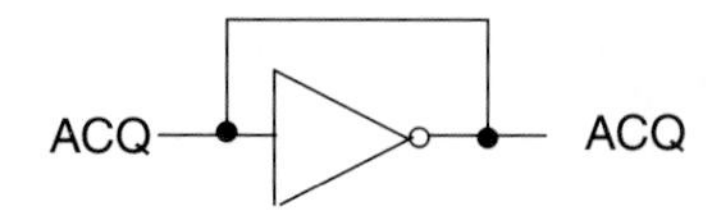

Figure 12-5 A clock generator.

Elements of a vector signal can also be accessed, as either a single element or a slice. For example,

```
signal A: STD_LOGIC;
signal C: STD_LOGIC_VECTOR(0 to 4);
signal B, D: STD_LOGIC_VECTOR(5 downto 0);
   . . .
D(4 downto 0) <= B(5 downto 1) or C;
D(5) <= A and B(5);
```

The first signal assignment implies:

```
D(4) <= B(5) or C(0);
D(3) <= B(4) or C(1);
. . .
```

Single-bit signals and vector signals can be concatenated to form larger vector signals. For example,

```
signal C, CC: STD_LOGIC_VECTOR(7 downto 0);
signal CX: STD_LOGIC;
. . .
C <= CX & CC(6 downto 0);
```

It is also possible to refer to an element of a vector signal whose index value is computable only at runtime. For example,

```
DOT <= DCR (SEL_ADDR);
```

implies a multiplexer whose output is signal DOT, and SEL_ADDR specifies the selection address. DCR is a vector of values of the same type as signal DOT. Object DCR models the behavior of the multiplexer. Here are some possible declarations for these.

```
signal SEL_ADDR: INTEGER range 0 to 255;
type MEM_TYPE is array (NATURAL range <>)
    of STD_LOGIC_VECTOR(0 to 3);
signal DCR: MEM_TYPE(0 to 255);
signal DOT: STD_LOGIC_VECTOR(0 to 3);
```

Shift operations can be performed using the predefined shift operators. Alternately, shift operations can be modeled using the concatenation operator. For example,

```
signal A, Z: STD_LOGIC_VECTOR(0 to 7);
. . .
Z <= A(1 to 7) & A(0);               -- A left rotate operation.
Z <= A(7) & A(0 to 6);               -- A right rotate operation.
Z <= A(1 to 7) & '0';                -- A left shift operation.
Z <= A(K to 7) & A(0 to (K–1));      -- A K-left rotate operation, K <= 7.
```

Subfields of a vector signal, called *slices*, can also be accessed. For example, consider a 32-bit instruction register, INSTR_REG, in which the first 16 bits denote the address, next 8 bits represent the opcode, and the remaining 8 bits represent the index. Given the following declarations,

```
type MEMORY_TYPE is
  array (0 to 1023) of STD_LOGIC_VECTOR(31 downto 0);
signal INSTR_REG: STD_LOGIC_VECTOR(31 downto 0);
signal ADDRESS: STD_LOGIC_VECTOR(15 downto 0);
signal OP_CODE, INDEX: STD_LOGIC_VECTOR(7 downto 0);
signal MEMORY: MEMORY_TYPE;
signal PROG_CTR: STD_LOGIC_VECTOR(0 to 9);
```

one way to read the subfield information from the INST
use three concurrent signal assignment statements.
instruction register are assigned to specific sig

```
INSTR_REG <= MEMORY (TO_INTEGER(PROG_CTR));
-- TO_INTEGER is a function that translates
-- STD_LOGIC_VECTOR value to integer (see package UTILS_PKG).
ADDRESS <= INSTR_REG(31 downto 16);
OP_CODE <= INSTR_REG(15 downto 8);
INDEX <= INSTR_REG(7 downto 0);
. . .
PROCEDURE_CALL(ADDRESS, OP_CODE, INDEX);
```

An alternative, more efficient way is to declare the subfields of the instruction register as aliases instead of as separate signals.

```
alias ADDRESS: STD_LOGIC_VECTOR(15 downto 0) is
  INSTR_REG(31 downto 16);
alias OP_CODE: STD_LOGIC_VECTOR(7 downto 0) is
  INSTR_REG(15 downto 8);
alias INDEX: STD_LOGIC_VECTOR(7 downto 0) is
  INSTR_REG(7 downto 0);
```

In this case, explicit assignments to the three subfields are not necessary, as shown next.

```
INSTR_REG <= MEMORY (TO_INTEGER(PROG_CTR));
. . .
PROCEDURE_CALL (ADDRESS, OP_CODE, INDEX);
```

12.3 Different Styles of Modeling

This section gives examples of the three different modeling styles provided by the language: dataflow, behavioral, and structural. Consider the circuit shown in Figure 12-6, which saves the value of the input A into a register and then multiplies it with input C.

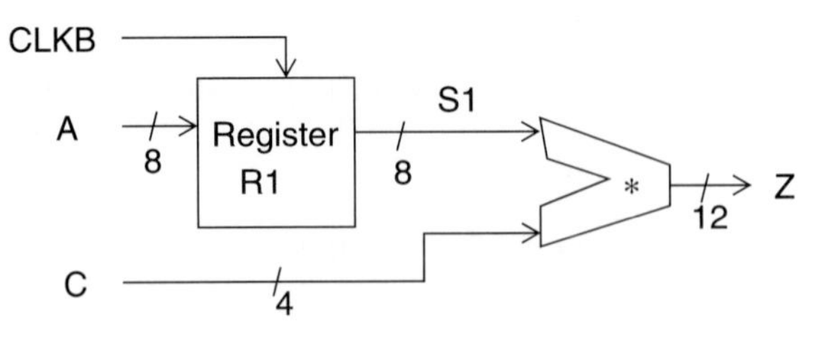

Figure 12-6 A buffered multiplier.

The entity declaration that declares the interface signals for the circuit is:

```
library IEEE, BOOK_LIB;
use IEEE.STD_LOGIC_1164.all, BOOK_LIB.UTILS_PKG.all;
entity SAVE_MULT is
  port ( A: in STD_LOGIC_VECTOR(0 to 7);
         C: in STD_LOGIC_VECTOR(0 to 3);
         CLKB: in STD_LOGIC;
         Z: out STD_LOGIC_VECTOR(0 to 11));
end SAVE_MULT;
```

The first modeling style is the dataflow style in which concurrent signal assignment statements are used to model the circuit.

```
architecture DATA_FLOW of SAVE_MULT is
  signal S1: STD_LOGIC_VECTOR(0 to 7);
begin
  Z <= S1 * C;        -- The multiplication operator is overloaded
                      -- (in package UTILS_PKG) to operate
                      -- on STD_LOGIC_VECTORs of different sizes.
  R1: block (CLKB = '1' and not CLKB'STABLE)
  begin
    S1 <= guarded A;
  end block R1;
end DATA_FLOW;
```

This representation does not directly imply any structure, but implicitly describes it. However, its functionality is very clear. The register has been modeled using a block statement.

The second way to describe the circuit is to model it as a sequential process.

```
architecture SEQUENTIAL of SAVE_MULT is
  signal S1: STD_LOGIC_VECTOR(0 to 7);
begin
  process (A, C, CLKB, S1)
  begin
    if CLKB = '1' and not CLKB'STABLE then
      S1 <= A;
    end if;
    Z <= S1 * C;
  end process;
end SEQUENTIAL;
```

This model also describes the behavior, but does not imply any structure, either explicitly or implicitly. In this case, the register has been modeled using an `if` statement.

The third way to describe the SAVE_MULT circuit is to model it as a netlist assuming the existence of an 8-bit register and an 8-bit multiplier.

```
architecture NET_LIST of SAVE_MULT is
  component REG8
    port ( DIN: in STD_LOGIC_VECTOR(0 to 7);
           CLK: in STD_LOGIC;
           DOUT: out STD_LOGIC_VECTOR(0 to 7));
  end component;
  component MULT8
    port ( A, B: in STD_LOGIC_VECTOR(0 to 7);
           Z: out STD_LOGIC_VECTOR(0 to 15));
  end component;
  signal S1, S3     : STD_LOGIC_VECTOR(0 to 7);
  signal S2         : STD_LOGIC_VECTOR(0 to 15);
begin
  R1: REG8 port map (A, CLKB, S1);
  M1: MULT8 port map (S1, S3, S2);
  Z <= S2 (4 to 15);
  S3 <= "0000" & C;        -- This assignment is necessary since the
    -- right-hand-side expression cannot be used in the port map
    -- directly since it is not a globally static expression.
end NET_LIST;
```

This description explicitly describes the structure, but the behavior is unknown. This is because the REG8 and MULT8 component names are arbitrary, and they could have any behavior associated with them. The signal assignment to Z is necessary because the multiplier was assumed to produce a 16-bit output.

Of these three different modeling styles, the behavioral style of modeling is generally the fastest to simulate.

12.4 Modeling Regular Structures

Let us assume that we want to construct a 10-deep, 8-bit stack using 8-bit registers. The stack has an input port, an output port, and a control signal that specifies whether to pop or to push data. Overflow and underflow conditions are ignored. Ten 8-bit registers model the stack. The input to each register comes from either the top register or the bottom register, depending on whether to push or pop. Since this basically involves a replication of the register 10 times, a generate statement can be used to model the stack. Here is the model.

```
library IEEE; use IEEE.STD_LOGIC_1164.all;
entity STACK is
  port (DATAIN: in STD_LOGIC_VECTOR(7 downto 0);
        DATAOUT: out STD_LOGIC_VECTOR(7 downto 0);
        CLK, STCK_CTRL: in STD_LOGIC);
end STACK;

architecture GENERATE_EX of STACK is
  type STACK_TYPE is array (1 to 10) of
                STD_LOGIC_VECTOR(7 downto 0);
  signal REGIN: STD_LOGIC_VECTOR(7 downto 0);
  signal DATA: STACK_TYPE;
  component REG8
    port ( DIN: in STD_LOGIC_VECTOR(0 to 7);
           DOUT: out STD_LOGIC_VECTOR(0 to 7);
           CLK: in STD_LOGIC);
  end component;
begin
  G1: for K in 1 to 10 generate          -- 10 is top of stack, 1 is bottom.
    G2: if K = 10 generate
      REGIN <=    DATAIN when STCK_CTRL = '1'
                  else DATA(K–1);
    end generate G2;
    G3: if K > 1 and K < 10 generate
      REGIN <=    DATA(K+1) when STCK_CTRL = '1'          -- PUSH
                  else DATA(K–1);                         -- POP
    end generate G3;
    G4: if K = 1 generate
      REGIN <=    DATA(K+1) when STCK_CTRL = '1'
                  else (others => 'U');
    end generate G4;
    FF8: REG8 port map (REGIN, DATA(K), CLK);
  end generate G1;
  DATAOUT <= DATA(10);          -- Output is connected to top of stack.
end GENERATE_EX;
```

When variable K has a value of 1 and the stack is being popped, the input to the bottom-most register is 'U'; when K is 10, the input to the top-most register is the input to the stack, DATAIN. The output of the top-most register is the output of the stack.

12.5 Modeling Delays

Consider a 3-input `nor` gate. Its behavior can be modeled using a concurrent signal assignment statement, such as:

```
-- A, B, C and Z are signals of type STD_LOGIC.
Z <= not (A or B or C) after 12 ns;
```

This statement models the `nor` gate with an inertial delay of 12 ns. This delay represents the time from an event on signal A, B, or C until the result value appears on signal Z. An event could be any value change, for example, 'U'->'Z', 'U'->'0', or '1'->'0'.

If the rise time, that is, the transition time for '0'->'1', and the fall time, that is, the transition time for '1'->'0', were to be explicitly modeled, a different mechanism would be needed. Consider the following conditional signal assignment.

```
signal Z, A: BIT;
. . .
Z <=   '0' after 12 ns when A = '1' else
       '1' after 14 ns;                          -- when A = '0';
```

If an event occurs on signal A (it can only be a '0'->'1' or a '1'->'0' transition since the signal's type is BIT), the value '0' will appear on signal Z after 12 ns if signal A had the transition '0'->'1' (12 ns is then the fall time); otherwise, Z will get the value '1' after 14 ns if the transition on signal A is from '1'->'0' (14 ns is then the rise time). Such a model is sufficient for BIT types but is insufficient when a signal type has more than two values, for example, when using an STD_LOGIC type. In such a case, the following example illustrates an approach to modeling rise and fall times independently.

```
library IEEE, BOOK_LIB;
use IEEE.STD_LOGIC_1164.all;
use BOOK_LIB.UTILS_PKG.all;
entity NOR3 is
  port (A, B, C: in STD_LOGIC; Z: out STD_LOGIC);
end NOR3;

architecture TIMES of NOR3 is
begin
  process (A, B, C)
    variable OLD_Z, NEW_Z: STD_LOGIC;
    constant RISE_TIME: TIME := 8 ns;
```

```
      constant FALL_TIME: TIME := 10 ns;
   begin
      NEW_Z := not (A or B or C);
      if OLD_Z = '0' and NEW_Z = '1' then
         Z <= '1' after RISE_TIME;
      elsif OLD_Z = '1' and NEW_Z = '0' then
         Z <= '0' after FALL_TIME;
      else
         Z <= NEW_Z after MAX(RISE_TIME, FALL_TIME);
         -- Function MAX returns the maximum of the two time
         -- values. It is declared in the package UTILS_PKG.
      end if;
      OLD_Z := NEW_Z;
   end process;
end TIMES;
```

There is nothing special in the previous example. The old value of signal Z is saved. When a new value for Z is computed, a check is made to see what type of transition occurred. Based on whether a rising or falling edge occurred, the computed value is assigned to the signal Z after the appropriate delay. The rise and fall times are specified using constant declarations in this example. They could also have been declared elsewhere, for example, in an entity declaration or in a package declaration or as generics for the NOR3 entity, such as:

```
entity NOR3 is
   generic (RISE_TIME: TIME := 8 ns; FALL_TIME: TIME := 10 ns);
   port (A, B, C: in STD_LOGIC; Z: out STD_LOGIC);
end NOR3;
```

12.6 Modeling Conditional Operations

Operations that occur under certain conditions can be modeled using either a selected signal assignment statement, a conditional signal assignment statement, an `if` statement, or a `case` statement. Let us consider a 3-bit decoder circuit. Its behavior can be modeled using a conditional signal assignment, as shown below.

```
library IEEE, BOOK_LIB;
use IEEE.STD_LOGIC_1164.all;
use BOOK_LIB.UTILS_PKG.all;
entity DECODER is
```

```
    port ( CTRL: in STD_LOGIC_VECTOR(0 to 2);
           Z: out STD_LOGIC_VECTOR(0 to 7));
    constant DECODER_DELAY: TIME := 24 ns;
end DECODER;

architecture CONDITIONAL_EX of DECODER is
begin
    G1: for K in 0 to 7 generate
        Z(K) <= '0' after DECODER_DELAY
                when TO_INTEGER(CTRL) = K else
                '1' after DECODER_DELAY;
        -- TO_INTEGER is a function declared in the package
        -- UTILS_PKG; it converts a STD_LOGIC_VECTOR value
        -- into an integer value.
    end generate G1;
end CONDITIONAL_EX;
```

A multiplexer can be modeled using a process statement. The value of the select lines are first determined and, based on this value, a `case` statement selects the appropriate input that is to be assigned to the output.

```
library IEEE; use IEEE.STD_LOGIC_1164.all;
entity MULTIPLEXER is
    generic (MUX_DELAY: TIME := 15 ns);
    port ( SEL: in STD_LOGIC_VECTOR(0 to 1);
           A, B, C, D: in STD_LOGIC;
           MUX_OUT: out STD_LOGIC);
end MULTIPLEXER;

architecture PROCESS_MUX of MULTIPLEXER is
begin
    P1: process (SEL, A, B, C, D)
        variable TEMP: STD_LOGIC;
    begin
        case SEL is
            when "00" => TEMP := A;
            when "01" => TEMP := B;
            when "10" => TEMP := C;
            when "11" => TEMP := D;
            when others => TEMP := '-';
        end case;
        MUX_OUT <= TEMP after MUX_DELAY;
    end process P1;
end PROCESS_MUX;
```

12.7 Modeling Synchronous Logic

So far in this chapter, most of the examples that we have seen are of combinational logic. As far as synchronous logic is concerned, there is no explicit object in the language that declares registers or memories. Instead, the semantics of the language provides for signals to be interpreted as registers or memories depending on how the values to these signals are assigned. There are again a great number of ways in which synchronous logic can be modeled. Some of these are:

- By controlling the signal assignment
- By using a guarded assignment
- By using a `register` kind signal

Consider the following example, which shows how controlling a signal assignment can model a synchronous, edge-triggered D-type flip-flop.

```
library IEEE; use IEEE.STD_LOGIC_1164.all;
entity D_FLIP_FLOP is
  port (D, CLOCK: in STD_LOGIC; Q: out STD_LOGIC);
end D_FLIP_FLOP;

architecture SYNCHRONOUS of D_FLIP_FLOP is
begin
  process (CLOCK)
  begin
    if CLOCK = '1' and CLOCK'EVENT then              -- Rising edge.
      Q <= D after 5 ns;
    end if;
  end process;
end SYNCHRONOUS;
```

The semantics of the process statement indicates that when there is a rising edge on signal CLOCK, Q will get the value of D after 5 ns, else the value of signal Q does not change (a signal retains its value until it is assigned a new value). Even though Q and D are signals, the behavior in the process statement expresses the semantics of a D-type flip-flop. Given this entity, an 8-bit register can be modeled as follows:

```
library IEEE; use IEEE.STD_LOGIC_1164.all;
entity REGISTER8 is
  generic (START: INTEGER := 0; STOP: INTEGER := 7);
```

```
   port ( D: in STD_LOGIC_VECTOR(START to STOP);
          Q: out STD_LOGIC_VECTOR(START to STOP);
          CLOCK: in STD_LOGIC);
end REGISTER8;

architecture COMPONENT_BEH of REGISTER8 is
   component D_FLIP_FLOP
      port (D, CLOCK: in STD_LOGIC; Q: out STD_LOGIC);
   end component;
begin
   -- Assumed START <= STOP.
   G_LABEL: for K in START to STOP generate
      DFF: D_FLIP_FLOP port map (D(K), CLOCK, Q(K));
      -- This component instantiation statement could very well have
      -- been replaced by the process statement that appears
      -- in the architecture body of the D_FLIP_FLOP entity.
   end generate G_LABEL;
end COMPONENT_BEH;
```

Consider a gated cross-coupled latch circuit, as shown in Figure 12-7, and its dataflow model.

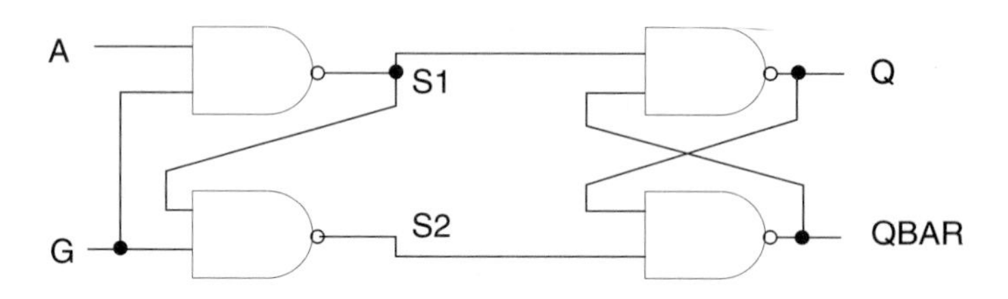

Figure 12-7 A gated latch.

```
library IEEE; use IEEE.STD_LOGIC_1164.all;
entity GATED_FF is
   port (A, G: in STD_LOGIC; Q, QBAR: buffer STD_LOGIC);
end GATED_FF;

architecture IMPLICIT_LATCH of GATED_FF is
   signal S1, S2: STD_LOGIC;
begin
   S1 <= A nand G;
   S2 <= S1 nand G;
   Q <= QBAR nand S1;
   QBAR <= Q nand S2;
end IMPLICIT_LATCH;
```

In this example, the concurrent signal assignment statements do not have any guards associated with them, but the semantics of the behavior implies a latch. Note that the outputs of the gated flip-flop are of mode **buffer** since they are assigned to and read within the architecture body.

A memory can be modeled as a two-dimensional array of flip-flops. Consider a generic RAM model in which each memory element is a GATED_FF component, described earlier. ASIZE is the number of bits on the address port, and DSIZE is the number of bits on the data port of the RAM.

```
library IEEE; use IEEE.STD_LOGIC_1164.all;
entity RAM_GENERIC is
  generic (ASIZE, DSIZE: POSITIVE);
  port ( ADDRESS: in STD_LOGIC_VECTOR(ASIZE–1 downto 0);
         DATA_IN: in STD_LOGIC_VECTOR(DSIZE–1 downto 0);
         DATA_OUT: buffer
                     STD_LOGIC_VECTOR(DSIZE–1 downto 0));
end RAM_GENERIC;

architecture COMPONENT_BEH of RAM_GENERIC is
  component GATED_FF
    port (A, G: in STD_LOGIC; Q, QBAR: buffer STD_LOGIC := '0');
  end component;
  component DECODER
    generic (SELECT_SIZE: INTEGER);
    port ( ADDR: out
             STD_LOGIC_VECTOR(2**SELECT_SIZE–1 downto 0);
           SEL_CTRL: in
             STD_LOGIC_VECTOR(SELECT_SIZE–1 downto 0));
  end component;
  signal ADDR_SELECT:
                     STD_LOGIC_VECTOR(2**ASIZE–1 downto 0);
begin
  D1: DECODER
        generic map (ASIZE)
        port map (ADDR_SELECT, ADDRESS);
  L1: for J in 0 to 2**ASIZE–1 generate
    L2: for K in 0 to DSIZE–1 generate
      GFF: GATED_FF port map (DATA_IN(K),
                        ADDR_SELECT(J), DATA_OUT(K), open);
    end generate L2;
  end generate L1;
end COMPONENT_BEH;
```

The keyword **open** in the instantiation of the gated latch indicates that the QBAR port of the gated latch is not connected to any signal.

Synchronous logic can also be modeled using guarded assignments. For example, a level-sensitive D flip-flop can be modeled as:

```
library IEEE; use IEEE.STD_LOGIC_1164.all;
entity LEVEL_SENS_DFF is
   port (STROBE, D: in STD_LOGIC; Q, QBAR: out STD_LOGIC);
end LEVEL_SENS_DFF;

architecture GUARDED_BEH of LEVEL_SENS_DFF is
   signal TEMP: STD_LOGIC;
begin
   B1: block (STROBE = '1')
         -- STROBE is the name of the clock input.
   begin
      TEMP <= guarded D;
      Q <= TEMP;
      QBAR <= not TEMP;
   end block B1;
end GUARDED_BEH;
```

When STROBE is '1', any events on signal D are transferred to TEMP and eventually to Q, but when STROBE becomes '0', the value in TEMP (also in Q and QBAR) is retained, and any change in input D no longer affects the value of signal TEMP.

It is important to understand the semantics of a signal assignment to determine the inference of synchronous logic. Consider the difference between the following two architecture bodies, BODY1 and BODY2.

```
architecture BODY1 of NO_NAME is
   signal A: STD_LOGIC;
begin
   A <= not A;
end BODY1;

architecture BODY2 of NO_NAME is
   signal A, CLOCK: STD_LOGIC;
begin
   process (A, CLOCK)
   begin
      if CLOCK = '0' and CLOCK'EVENT then
         A <= not A;
      end if;
   end process;
end BODY2;
```

Architecture BODY1 implies the circuit shown in Figure 12-8, while architecture BODY2 implies the circuit shown in Figure 12-9.

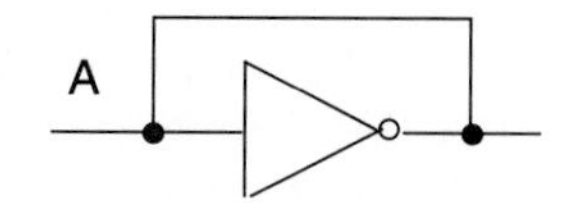

Figure 12-8 No latch implied.

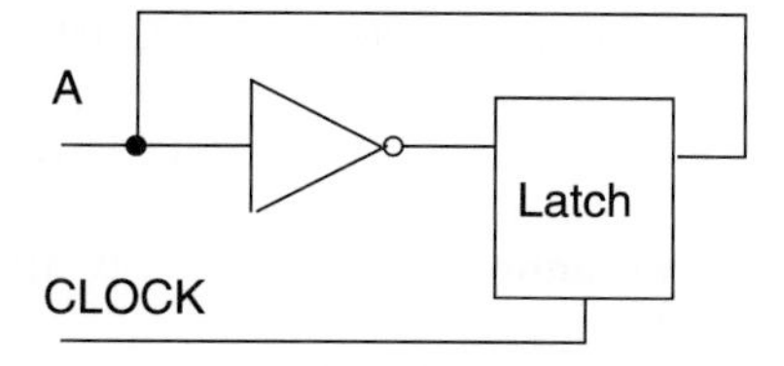

Figure 12-9 A latch implied.

If architecture BODY1 were simulated as is, simulation would go into an endless loop due to the delta delay asynchronous loop (simulation time does not advance), and each execution of the concurrent signal assignment statement would keep triggering itself after a delta delay. In architecture BODY2, the value of A is latched only on the falling edge of the CLOCK signal, and thereafter, any changes on A (input of latch) do not affect the output of latch.

A signal declared to be of the `register` kind also models a latch. If all drivers to such a signal are disconnected, the signal retains its old value, thereby implying a latch. Consider the circuit shown in Figure 12-10, in which a capacitive latch is loaded with different values based on the two mutually exclusive clock signals. The disconnect time models the decay time for the capacitances C1 and C2 that appear at the input of the drivers for signal S1.

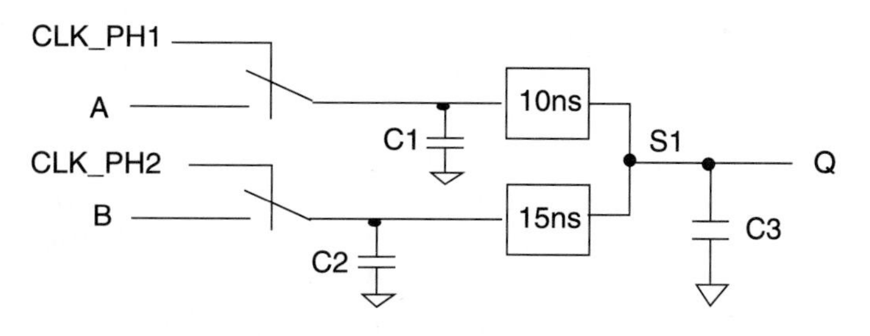

Figure 12-10 A capacitive latch.

```
library IEEE, BOOK_LIB;
use IEEE.STD_LOGIC_1164.all;
use BOOK_LIB.UTILS_PKG.all;
entity REGISTER_SIGNAL is
   port ( A, B, CLK_PH1, CLK_PH2: in STD_LOGIC;
            Q: out STD_LOGIC);
end REGISTER_SIGNAL;

architecture EXAMPLE of REGISTER_SIGNAL is
   signal S1: WIRED_OR STD_ULOGIC register;
      -- STD_ULOGIC is an unresolved 9-value type defined in
      -- package STD_LOGIC_1164. Function WIRED_OR is defined
      -- in package UTILS_PKG.
   disconnect S1: STD_ULOGIC after 32 ns;
begin
   B1: block (CLK_PH1 = '1')
   begin
      S1 <= guarded A after 10 ns;
   end block B1;
   B2: block (CLK_PH2 = '1')
   begin
      S1 <= guarded B after 15 ns;
   end block B2;
   Q <= S1;
end EXAMPLE;
```

12.8 State Machine Modeling

State machines can usually be modeled using a `case` statement in a process. The state information is stored in a signal. The multiple branches of the `case` statement contain the behavior for each state. Here is an example of a simple multiplication algorithm represented as a state machine. When the RESET signal is high, the accumulator ACC and the counter COUNT are initialized. When RESET goes low, multiplication starts. If the bit of the multiplier MPLR in position COUNT is '1', the multiplicand MCND is added to the accumulator. Next, the multiplicand is left-shifted by one bit, and the counter is incremented. If COUNT is 16, multiplication is complete, and the DONE signal is set high. If not, the COUNT bit of the multiplier MPLR is checked and the process repeated. The state diagram is shown in Figure 12-11, and the corresponding state machine model is shown next.

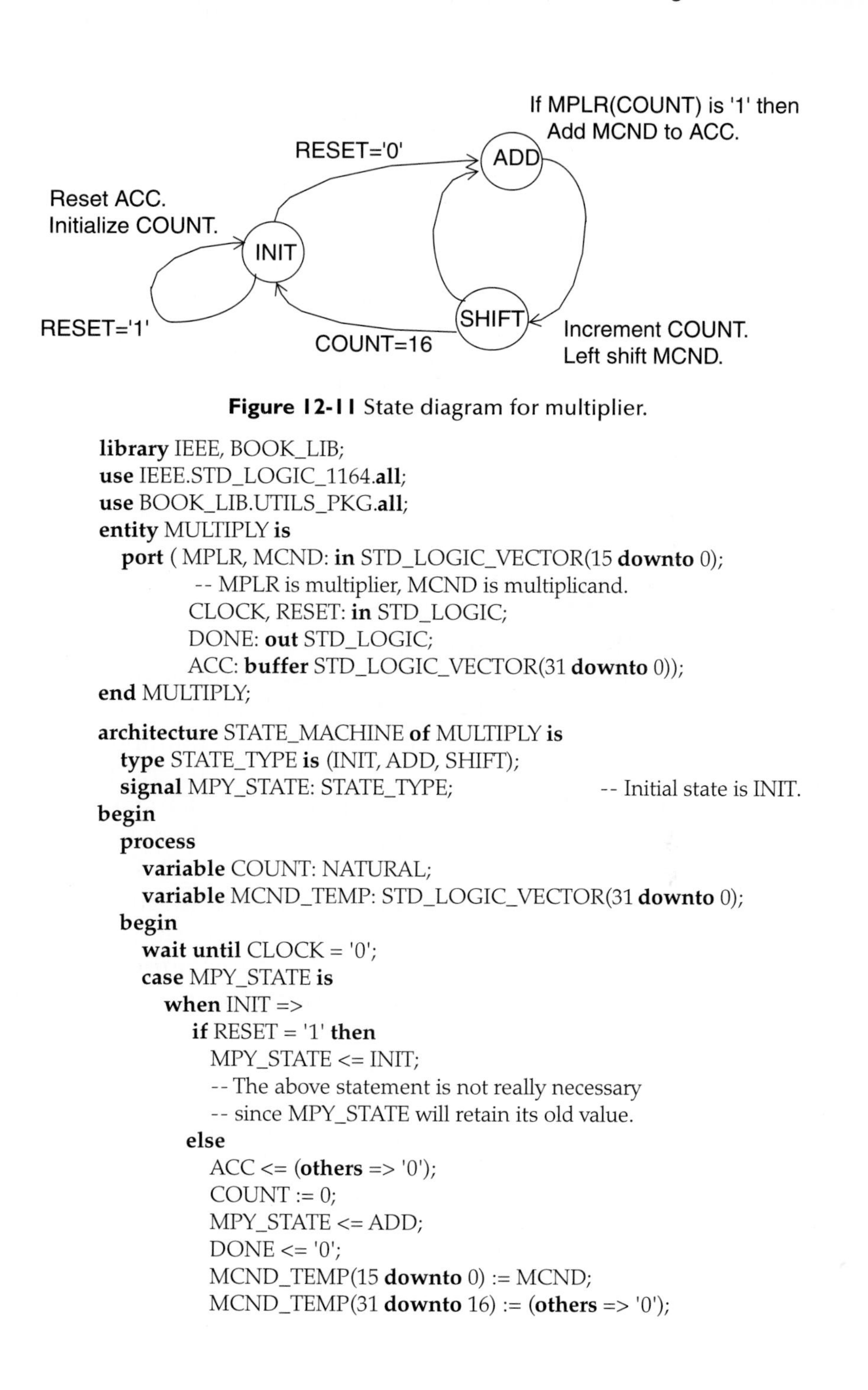

Figure 12-11 State diagram for multiplier.

```
library IEEE, BOOK_LIB;
use IEEE.STD_LOGIC_1164.all;
use BOOK_LIB.UTILS_PKG.all;
entity MULTIPLY is
  port ( MPLR, MCND: in STD_LOGIC_VECTOR(15 downto 0);
          -- MPLR is multiplier, MCND is multiplicand.
         CLOCK, RESET: in STD_LOGIC;
         DONE: out STD_LOGIC;
         ACC: buffer STD_LOGIC_VECTOR(31 downto 0));
end MULTIPLY;

architecture STATE_MACHINE of MULTIPLY is
  type STATE_TYPE is (INIT, ADD, SHIFT);
  signal MPY_STATE: STATE_TYPE;                    -- Initial state is INIT.
begin
  process
    variable COUNT: NATURAL;
    variable MCND_TEMP: STD_LOGIC_VECTOR(31 downto 0);
  begin
    wait until CLOCK = '0';
    case MPY_STATE is
      when INIT =>
        if RESET = '1' then
          MPY_STATE <= INIT;
          -- The above statement is not really necessary
          -- since MPY_STATE will retain its old value.
        else
          ACC <= (others => '0');
          COUNT := 0;
          MPY_STATE <= ADD;
          DONE <= '0';
          MCND_TEMP(15 downto 0) := MCND;
          MCND_TEMP(31 downto 16) := (others => '0');
```

```
            end if;
          when ADD =>
            if MPLR(COUNT) = '1' then
              ACC <= ACC + MCND_TEMP;
              -- The"+"overloaded operator is defined in
              -- package UTILS_PKG.
            end if;
            MPY_STATE <= SHIFT;
          when SHIFT =>
            -- Left-shift MCND:
            MCND_TEMP := MCND_TEMP(30 downto 0) & '0';
            COUNT := COUNT + 1;
            if COUNT = 16 then
              MPY_STATE <= INIT;
              DONE <= '1';
            else
              MPY_STATE <= ADD;
            end if;
        end case;
      end process;
    end STATE_MACHINE;
```

The signal MPY_STATE holds the state of the model. Initially, the model is in state INIT, and it stays in this state as long as signal RESET is '1'. When RESET gets the value '0', the accumulator ACC is cleared, the counter COUNT is reset, and the multiplicand MCND is loaded into a temporary variable MCND_TEMP, and the model advances to state ADD. When model is in the ADD state, the multiplicand in MCND_TEMP is added to ACC only if the bit at the COUNT position of the multiplier is a '1', and then the model advances to state SHIFT. In this state, the multiplier is left-shifted once, the counter is incremented, and, if the counter value is 16, signal DONE is set to '1' and the model returns to state INIT. At this time, ACC contains the result of the multiplication. If the counter value was less than 16, the model repeats itself going through states ADD and SHIFT until the counter value becomes 16.

State transitions occur at every falling clock edge; this is specified by the `wait` statement. The mode of signal ACC is set to **buffer** since the value is read and updated within the model.

12.9 Interacting State Machines

Interacting state machines can be described as separate processes communicating via common signals. Consider the state transition diagram shown in Figure 12-12 for two interacting processes, TX, a transmitter, and MP, a microprocessor. If process TX is not busy, process MP sets the data to be transmitted on a data bus and sends a signal, LOAD_TX, to process TX to load the data and begin transmitting. A signal, TX_BUSY, is set by process TX during transmission to indicate that it is busy and cannot receive any further data from process MP.

A skeleton model for these two interacting processes is shown. Only the control signals and state transitions are shown. Data manipulation code is not described.

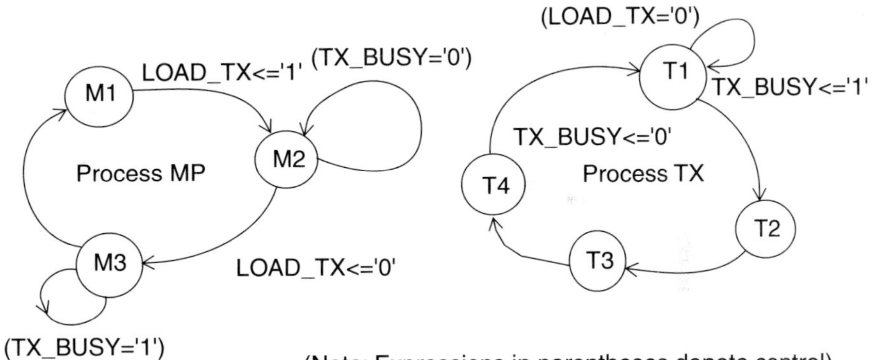

Figure 12-12 State diagram of two interacting processes.

```
architecture INTERACTING of FSM1 is
  type MP_STATE_TYPE is (M1, M2, M3);
  type TX_STATE_TYPE is (T1, T2, T3, T4);
  signal MP_STATE: MP_STATE_TYPE;
  signal TX_STATE: TX_STATE_TYPE;
  signal LOAD_TX, TX_BUSY: BIT;
  -- CLOCK is an input signal.
begin
  MP: process
  begin
    wait until CLOCK = '0';
```

```
    case MP_STATE is
      when M1 =>                          -- Load data on data bus.
        LOAD_TX <= '1';
        MP_STATE <= M2;
      when M2 =>                          -- Wait for acknowledge.
        if TX_BUSY = '1' then
          MP_STATE <= M3;
          LOAD_TX <= '0';
        end if;
      when M3 =>                          -- Wait for TX to finish.
        if TX_BUSY = '0' then
          MP_STATE <= M1;
        end if;
    end case;
  end process MP;
  TX: process (CLOCK)
  begin
    if CLOCK = '0' then               --"and CLOCK'EVENT"is implied.
      case TX_STATE is
        when T1 =>                      -- Wait for data to load.
          if LOAD_TX = '1' then
            TX_STATE <= T2;
            TX_BUSY <= '1';             -- Read data from data bus.
          end if;
        when T2 =>                      -- Transmitting data.
          TX_STATE <= T3;
        when T3 =>
          TX_STATE <= T4;
        when T4 =>                      -- Transmission completed.
          TX_BUSY <= '0';
          TX_STATE <= T1;
      end case;
    end if;
  end process TX;
end INTERACTING;
```

Two different ways of modeling a falling clock edge are shown in the two processes, MP and TX. Both are equivalent in this case, since all statements in the processes are synchronized to the clock edge. However, if there were some asynchronous statements, it would be necessary to use the `if` statement style. The `wait` statement style could also be used, but then two separate processes would have to be used, one for the synchronous part and one for the asynchronous part. The sequence of actions for the previous entity is shown in Figure 12-13.

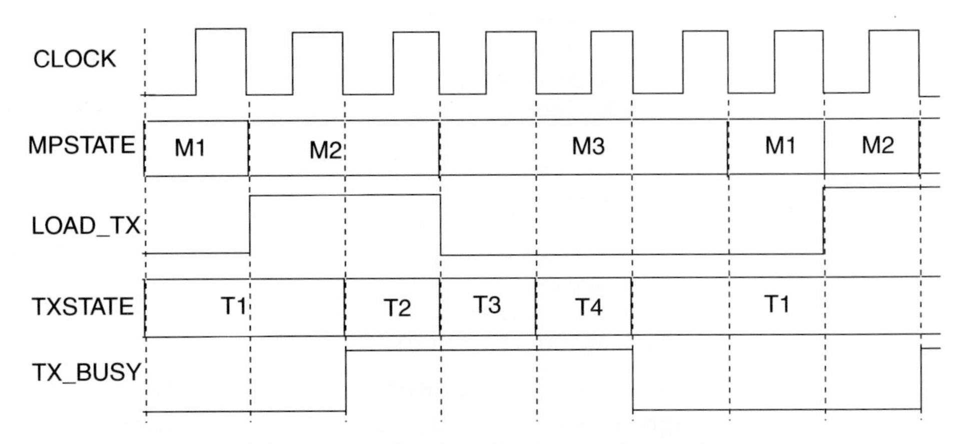

Figure 12-13 Sequence of actions for the two interacting processes.

Consider another example of two interacting processes, DIV, a clock divider, and RX, a receiver. In this case, process DIV generates a new clock, and process RX goes through its sequence of states synchronized to this new clock. The state diagram is shown in Figure 12-14.

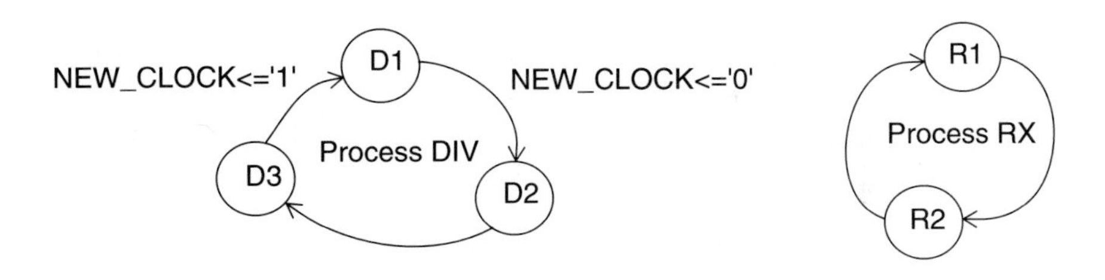

Figure 12-14 DIV generates clock for RX.

```
architecture ANOTHER_EXAMPLE of FSM2 is
  type DIV_STATE_TYPE is (D1, D2, D3);
  type RX_STATE_TYPE is (R1, R2);
  signal DIV_STATE: DIV_STATE_TYPE;
  signal RX_STATE: RX_STATE_TYPE;
  signal NEW_CLOCK: BIT;
  -- Signal CLOCK is an input port.
begin
  DIV: process
  begin
    wait until CLOCK = '1';
    case DIV_STATE is
      when D1 => DIV_STATE <= D2; NEW_CLOCK <= '0';
```

```
            when D2 => DIV_STATE <= D3;
            when D3 => NEW_CLOCK <= '1'; DIV_STATE <= D1;
        end case;
    end process DIV;
    RX: process
    begin
        wait until NEW_CLOCK = '0';
        case RX_STATE is
            when R1 => RX_STATE <= R2;
            when R2 => RX_STATE <= R1;
        end case;
    end process RX;
end architecture ANOTHER_EXAMPLE;
```

Process DIV generates a new clock, NEW_CLOCK, as it goes through its sequence of states. The state transitions in this process occur on the rising edge of the signal CLOCK. Process RX is executed every time a falling edge on the NEW_CLOCK signal occurs. The sequence of waveforms for these interacting state machines is shown in Figure 12-15.

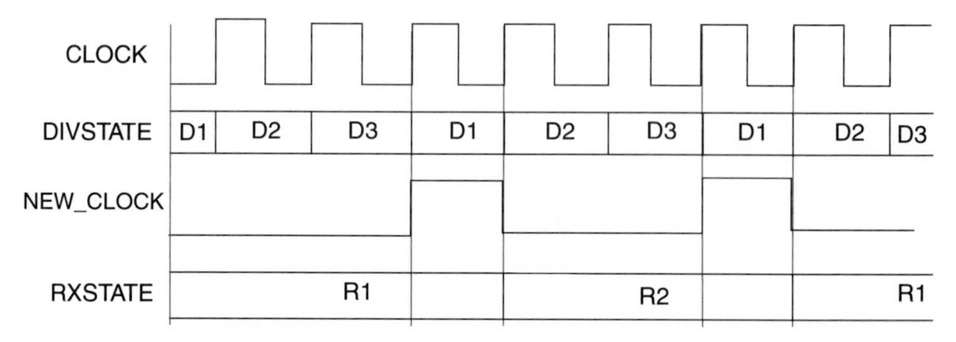

Figure 12-15 Interaction between processes, RX and DIV.

12.10 Modeling a Moore FSM

The output of a Moore finite state machine (FSM) depends only on the state and not on its inputs. This type of behavior can be modeled using a single process with a `case` statement that switches on the state value. An example of a state transition diagram for a Moore finite state machine is shown in Figure 12-16, and its corresponding behavior model appears next.

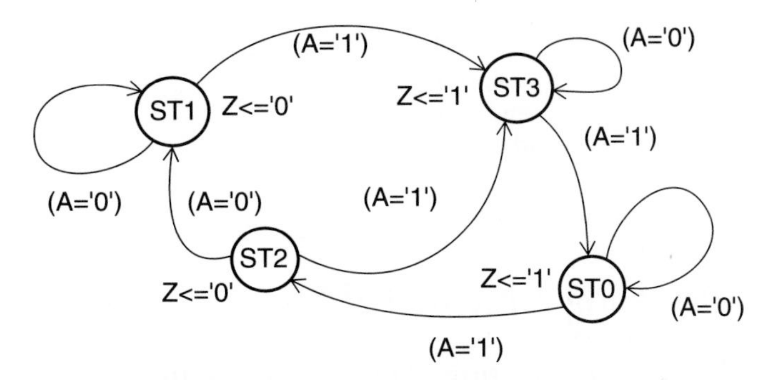

Figure 12-16 State diagram of a Moore machine.

```
library IEEE; use IEEE.STD_LOGIC_1164.all;
entity MOORE_FSM is
  port (A, CLOCK: in BIT; Z: out STD_LOGIC);
end MOORE_FSM;

architecture FSM_EXAMPLE of MOORE_FSM is
  type STATE_TYPE is (ST0, ST1, ST2, ST3);
  signal MOORE_STATE: STATE_TYPE;
begin
  process (CLOCK)
  begin
    if CLOCK = '0' then                              -- Falling edge.
      case MOORE_STATE is
        when ST0 =>
          Z <= '1';
          if A = '1' then
            MOORE_STATE <= ST2;
          end if;
        when ST1 =>
          Z <= '0';
          if A = '1' then
            MOORE_STATE <= ST3;
          end if;
        when ST2 =>
          Z <= '0';
          if A = '0' then
            MOORE_STATE <= ST1;
          else
            MOORE_STATE <= ST3;
```

```
        end if;
      when ST3 =>
        Z <= '1';
        if A = '1' then
          MOORE_STATE <= ST0;
        end if;
    end case;
  end if;
 end process;
end FSM_EXAMPLE;
```

12.11 Modeling a Mealy FSM

In a Mealy finite state machine, the outputs not only depend on the state of the machine but also on its inputs. This type of finite state machine can also be modeled in a style similar to that of the Moore example case, that is, using a single process. To show the variety of the language, a different style is used to model a Mealy machine. In this case we use two processes, one process that models the synchronous aspect of the finite state machine, and one that models the combinational part of the finite state machine. Here is an example of a state transition table, shown in Figure 12-17, and its corresponding behavior model.

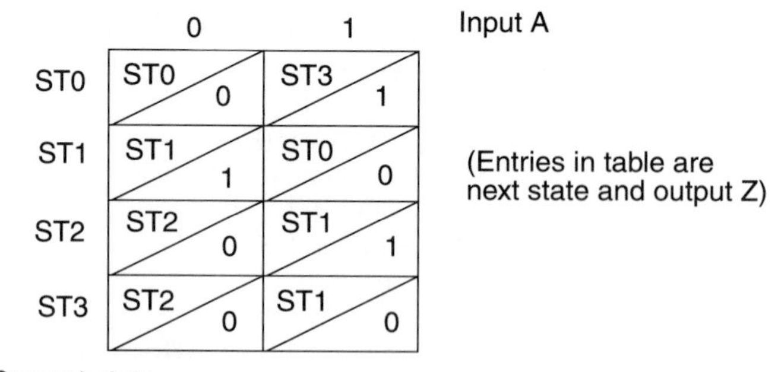

Figure 12-17 State transition table for a Mealy machine.

```
library IEEE; use IEEE.STD_LOGIC_1164.all;
entity MEALY_FSM is
  port (A, CLOCK: in BIT; Z: out STD_LOGIC);
end MEALY_FSM;
```

```
architecture YET_ANOTHER_EXAMPLE of MEALY_FSM is
  type MEALY_TYPE is (ST0, ST1, ST2, ST3);
  signal P_STATE, N_STATE: MEALY_TYPE;
begin
  SEQ_PART: process (CLOCK)
  begin
    -- Synchronous section:
    if CLOCK = '0' then                           -- Falling edge.
      P_STATE <= N_STATE;
    end if;
  end process SEQ_PART;

  COMB_PART: process (P_STATE, A)
  begin
    case P_STATE is
      when ST0 =>
        if A = '1' then
          Z <= '1'; N_STATE <= ST3;
        else
          Z <= '0';
        end if;
      when ST1 =>
        if A = '1' then
          Z <= '0'; N_STATE <= ST0;
        else
          Z <= '1';
        end if;
      when ST2 =>
        if A = '0' then
          Z <= '0';
        else
          Z <= '1'; N_STATE <= ST1;
        end if;
      when ST3 =>
        Z <= '0';
        if A = '0' then
          N_STATE <= ST2;
        else
          N_STATE <= ST1;
        end if;
    end case;
  end process COMB_PART;
end YET_ANOTHER_EXAMPLE;
```

In this type of finite state machine, it is important to put the input signals in the sensitivity list for the combinational part process, since the outputs may directly depend on the inputs independent of the clock. Such a condition does not occur in a Moore finite state machine since outputs depend only on states and state changes occur synchronously.

12.12 A Generic Priority Encoder

Here is a model of a generic priority encoder. Depending on which bit is a '1' at the input, the appropriate address appears on the output. Assume that the leftmost bit of the input vector DATA has the highest priority. Signal VALID is set to FALSE if none of the bits in the input vector are active (high).

```
library IEEE;
use IEEE.STD_LOGIC_1164.all;
entity PRIORITY_ENCODER is
  generic (BITS: NATURAL := 8);
  port (DATA: in STD_LOGIC_VECTOR(BITS downto 1);
        ADDR: out INTEGER range 0 to BITS–1;
        VALID: out BOOLEAN);
end PRIORITY_ENCODER;

architecture BEHAVIOR of PRIORITY_ENCODER is
begin
  process (DATA)
    variable BIT_POS: NATURAL;
  begin
    BIT_POS := 0;
    for K in DATA'RANGE loop
      BIT_POS := K;
      if DATA(K) = '1' then
        exit;                                    -- Exit loop
      end if;
    end loop;

    -- BIT_POS is 0 if no bit in DATA is a '1'.
    if BIT_POS = 0 then
      VALID <= FALSE;
    else
      VALID <= TRUE;
    end if;
    ADDR <= BIT_POS;
  end process;
end;
```

12.13 A Simplified Blackjack Program

This section presents a state machine description of a simplified blackjack program. The blackjack program is played with a deck of cards. Cards 2 to 10 have values equal to their face value, and an ace has a value of either 1 or 11. The object of the game is to accept a number of random cards such that the total score (sum of values of all cards) is as close as possible to 21 without exceeding 21.

When a new card is inserted, signal CARD_RDY is true, and signal CARD_VALUE has the value of the card. The signal REQUEST_CARD indicates when the program is ready to accept a new card. If a sequence of cards is accepted such that the total exceeds 21, signal LOST is set to true, indicating that it has lost; otherwise, the signal WON is set to true, indicating that the game has been won. The state sequencing is controlled by the signal CLOCK. This external interface of the blackjack program, shown in Figure 12-18, is expressed using the following entity declaration.

```
entity BLACKJACK is
  port (CARD_RDY: in BOOLEAN; CARD_VALUE: in INTEGER;
        REQUEST_CARD, WON, LOST: out BOOLEAN;
        CLOCK: in BIT);
end BLACKJACK;
```

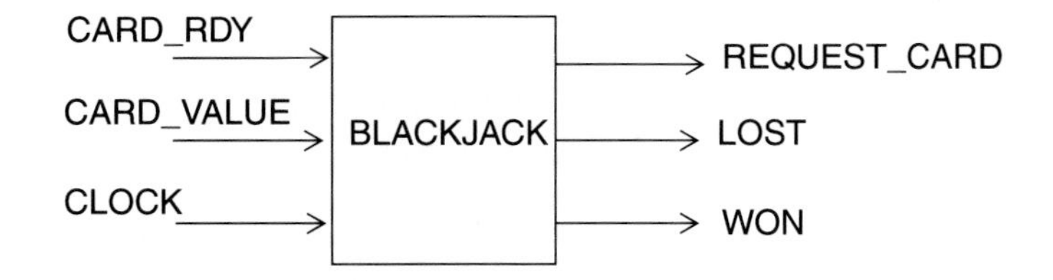

Figure 12-18 External view of blackjack program.

The behavior of the program is described in the following architecture body. The program accepts cards until its score is at least 17. The first ace is counted as an 11 unless the score exceeds 21, in which case 10 is subtracted so that the value of 1 is used for an ace. Three registers are used to store the state of the program: TOTAL to hold the sum, CURRENT_CARD_VALUE to hold the value of the card read (which could be 1 through 10), and ACE_AS_11 to remember whether an ace was counted as an 11 instead of a 1. The state of the blackjack program is stored in signal BJ_STATE.

```
architecture STATE_MACHINE of BLACKJACK is
  type STATE_TYPE is (INITIAL_ST, GETCARD_ST, REMCARD_ST,
      ADD_ST, CHECK_ST, WIN_ST, BACKUP_ST, LOSE_ST);
  signal BJ_STATE: STATE_TYPE;
begin
  process
    variable CURRENT_CARD_VALUE, TOTAL: INTEGER;
    variable ACE_AS_11: BOOLEAN;
  begin
    wait until CLOCK = '0';
    case BJ_STATE is
      when INITIAL_ST =>
        TOTAL := 0; ACE_AS_11 := FALSE; WON <= FALSE;
        LOST <= FALSE; BJ_STATE <= GETCARD_ST;
      when GETCARD_ST =>
        REQUEST_CARD <= TRUE;
        if CARD_RDY then
          CURRENT_CARD_VALUE := CARD_VALUE;
          BJ_STATE <= REMCARD_ST;
        end if;                        -- Else stay in GETCARD_ST state.
      when REMCARD_ST =>               -- Wait for card to be removed.
        if CARD_RDY then
          REQUEST_CARD <= FALSE;
        else
          BJ_STATE <= ADD_ST;
        end if;
      when ADD_ST =>
        if not ACE_AS_11 and
            CURRENT_CARD_VALUE = 1 then
          CURRENT_CARD_VALUE := 11;
          ACE_AS_11 := TRUE;
        end if;
        TOTAL := TOTAL + CURRENT_CARD_VALUE;
        BJ_STATE <= CHECK_ST;
      when CHECK_ST =>
        if TOTAL < 17 then
          BJ_STATE <= GETCARD_ST;
        elsif TOTAL < 22 then
          BJ_STATE <= WIN_ST;
        else
          BJ_STATE <= BACKUP_ST;
        end if;
      when BACKUP_ST =>
        if ACE_AS_11 then
```

```
          TOTAL := TOTAL – 10; ACE_AS_11 := FALSE;
          BJ_STATE <= CHECK_ST;
        else
          BJ_STATE <= LOSE_ST;
        end if;
      when LOSE_ST =>
        LOST <= TRUE; REQUEST_CARD <= TRUE;
        if CARD_RDY then
          BJ_STATE <= INITIAL_ST;
        end if;                        -- Else stay in this state.
      when WIN_ST =>
        WON <= TRUE; REQUEST_CARD <= TRUE;
        if CARD_RDY then
          BJ_STATE <= INITIAL_ST;
        end if;                        -- Else stay in this state.
    end case;
  end process;
end STATE_MACHINE;
```

12.14 A Clock Divider

A generic model of a divide-by-2*N clock generator is described.

```
library IEEE;
use IEEE.STD_LOGIC_1164.all;
entity CLOCK_DIVIDER is
  generic (N: POSITIVE);
  port (CLK, RST: in STD_LOGIC; CLK_DIV: buffer STD_LOGIC);
end CLOCK_DIVIDER;

architecture BEHAVIOR of CLOCK_DIVIDER is
begin
  process (CLK, RST)
    variable COUNT: NATURAL;
  begin
    if RST = '0' then
      COUNT := 0;
      CLK_DIV <= '0';
    elsif CLK'EVENT and CLK = '1' then
      COUNT := COUNT + 1;

      if COUNT = N then
        CLK_DIV <= not CLK_DIV;
        COUNT := 0;
```

```
        end if;
      end if;
    end process;
  end BEHAVIOR;
```

As long as RST is '0', the clock divider output stays at '0'. If RST goes to '1', the counter starts counting from the next rising clock edge. When N edges have been detected, the clock divider output is inverted and counter is reset to 0 and the cycle repeats. Note that all edges of the generated clock change only on the positive edge of the input clock. The waveform for a divide-by-6 clock divider is shown in Figure 12-19.

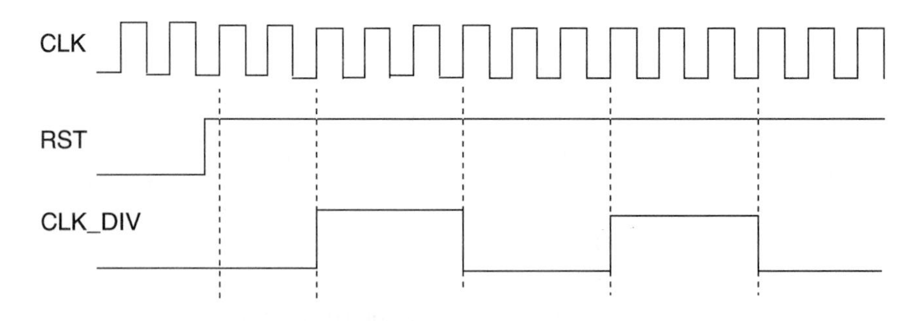

Figure 12-19 A clock divider.

12.15 A Generic Binary Multiplier

Here is a model of a generic N-by-M binary multiplier, N is the number of multiplier bits and M is the number of multiplicand bits.

```
-- This is a model for a 1-bit full-adder:
library IEEE;
use IEEE.STD_LOGIC_1164.all;
entity FULL_ADDER is
   port (X,Y, Z: in STD_LOGIC; S, C: out STD_LOGIC);
end FULL_ADDER;

architecture DF of FULL_ADDER is
begin
   S <= (X xorY) xor Z;
   C <= (X andY) or (X and Z) or (Y and Z);
end DF;
```

```
-- This is a model for a N-bit full-adder without carry-in:
library IEEE;
use IEEE.STD_LOGIC_1164.all;
entity GENERIC_FA is
  generic (N: NATURAL);
  port (A, B: in STD_LOGIC_VECTOR (N–1 downto 0);
        SUM: out STD_LOGIC_VECTOR (N–1 downto 0);
        COUT: out STD_LOGIC);
end GENERIC_FA;

architecture CONC of GENERIC_FA is
  component FULL_ADDER
    port (X,Y, Z: in STD_LOGIC; S, C: out STD_LOGIC);
  end component;
  signal CARRY: STD_LOGIC_VECTOR (N downto 0);
begin
  CARRY(0) <= '0';
  COUT <= CARRY(N);

  G1: for K in 0 to N–1 generate
    FA: FULL_ADDER port map (X => A(K),Y => B(K),
                Z => CARRY(K), S => SUM(K), C => CARRY(K+1));
  end generate;
end CONC;

-- This is the model for the binary multiplier:
library IEEE;
use IEEE.STD_LOGIC_1164.all;
entity BINARY_MULTIPLIER is
  generic (N, M: POSITIVE);
  port ( MPLR: in STD_LOGIC_VECTOR (N–1 downto 0);
        MPCD: in STD_LOGIC_VECTOR (M–1 downto 0);
        RESULT: out STD_LOGIC_VECTOR (N+M–1 downto 0));
end BINARY_MULTIPLIER;

architecture COMPACT of BINARY_MULTIPLIER is
  signal ACARRY: STD_LOGIC_VECTOR (N–1 downto 0);
  type SUM_TYPE is array (N–1 downto 0) of
      STD_LOGIC_VECTOR (M–1 downto 0);
  signal ASUM: SUM_TYPE;
  signal OPD1, OPD2: SUM_TYPE;
  signal RES: STD_LOGIC_VECTOR (N+M–1 downto 0);

  function RESIZE(A: in STD_LOGIC; SIZE: in NATURAL)
      return STD_LOGIC_VECTOR is
    variable RES: STD_LOGIC_VECTOR (SIZE–1 downto 0);
  begin
    RES := (others => A);
```

```
      return (RES);
   end;

   component GENERIC_FA
      generic (N: NATURAL);
      port (A, B: in STD_LOGIC_VECTOR (N–1 downto 0);
               SUM: out STD_LOGIC_VECTOR(N–1 downto 0);
               COUT: out STD_LOGIC);
   end component;
begin
   G2: for K in 1 to N–1 generate
      G3: if K = 1 generate
         ASUM(0) <= RESIZE(MPLR(0), M) and MPCD;
         RESULT(0) <= ASUM(0)(0);
         ACARRY(0) <= '0';
      end generate;

      OPD2(K) <= ACARRY(K-1) & ASUM(K–1)(M–1 downto 1);
      OPD1(K) <= RESIZE(MPLR(K), M) and MPCD;

      GFA: GENERIC_FA
         generic map (N => M)
         port map (A => OPD1(K), B => OPD2(K), SUM => ASUM(K),
                        COUT => ACARRY(K));

      RESULT(K) <= ASUM(K)(0);
   end generate;

   RESULT(N+M–1 downto M–1) <=
         ACARRY(N–1) & ASUM(N–1)(M–1 downto 1);
end COMPACT;
```

This is a standard binary multiplier. Each bit of the multiplier MPLR is multiplied with the multiplicand MPCD and a partial product is created. Each partial product is shifted left once before adding to create a partial sum.

Test Bench

Here is a test bench the tests a 3-bit by 4-bit binary multiplier. Generics have been used to represent the sizes so that the test bench can be customized for any other size multiplier. The multiplier and multiplicand values are provided on standard input (from keyboard) and the result appears on standard output (terminal screen).

```
library BOOK_LIB, IEEE;
use STD.TEXTIO.all; -- Includes TEXTIO stuff (READ, WRITE, LINE)
use BOOK_LIB.UTILS_PKG.all;                -- Includes functions used.
use IEEE.STD_LOGIC_1164.all;
entity TEST_BENCH is end;

architecture INTERACTIVE of TEST_BENCH is
  constant MPLR_SIZE: POSITIVE := 3;
  constant MPCD_SIZE: POSITIVE := 4;

  component BINARY_MULTIPLIER
    generic (N, M: POSITIVE);
    port (MPLR: in STD_LOGIC_VECTOR (N–1 downto 0);
          MPCD: in STD_LOGIC_VECTOR (M–1 downto 0);
          RESULT: out STD_LOGIC_VECTOR (N+M–1 downto 0));
  end component;

  signal LRSIG: STD_LOGIC_VECTOR (MPLR_SIZE–1 downto 0);
  signal CDSIG: STD_LOGIC_VECTOR (MPCD_SIZE–1 downto 0);
  signal RSIG: STD_LOGIC_VECTOR
                    (MPLR_SIZE + MPCD_SIZE – 1 downto 0);
begin
  -- Instantiate component under test:
  B1: BINARY_MULTIPLIER
    generic map (N => MPLR_SIZE, M => MPCD_SIZE)
    port map (MPLR => LRSIG,
              MPCD => CDSIG,
              RESULT => RSIG);

  process
    variable INBUF, OUTBUF: LINE;
    variable OPD1_STR: STRING (1 to MPLR_SIZE);
    variable OPD2_STR: STRING (1 to MPCD_SIZE);
  begin
    -- Read from standard input:
    READLINE (INPUT, INBUF);
    READ (INBUF, OPD1_STR);
    LRSIG <= TO_STDLOGICVECTOR (OPD1_STR);
      -- Function TO_STDLOGICVECTOR is defined in
      -- package UTILS_PKG.
    READ (INBUF, OPD2_STR);
    CDSIG <= TO_STDLOGICVECTOR (OPD2_STR);
    wait for 5 ns;

    -- Write the values to standard output:
    WRITE (OUTBUF, STRING'("MPLR="));
    WRITE (OUTBUF, OPD1_STR);
```

```
      WRITE (OUTBUF, STRING'("  MCND="));
      WRITE (OUTBUF, OPD2_STR);
      WRITE (OUTBUF, STRING'("  RESULT="));
      WRITE (OUTBUF, TO_STRING(RSIG));
        -- Function TO_STRING is defined in package UTILS_PKG.
      WRITELINE (OUTPUT, OUTBUF);
    end process;
  end;
```

Here is the input(I) and output(O) produced by the test bench.

```
I > 1000100
O > MPLR=100  MCND=0100  RESULT=0010000
I > 0011110
O > MPLR=001  MCND=1110  RESULT=0001110
I > 1111111
O > MPLR=111  MCND=1111  RESULT=1101001
I > 0010011
O > MPLR=001  MCND=0011  RESULT=0000011
```

12.16 A Pulse Counter

Here is an example of a circuit that counts the number of clock pulses (positive clock edges) contained within a signal SAMPLE. If count exceeds a MAX_COUNT, an overflow flag is set. Figure 12-20 shows the intention of this circuit.

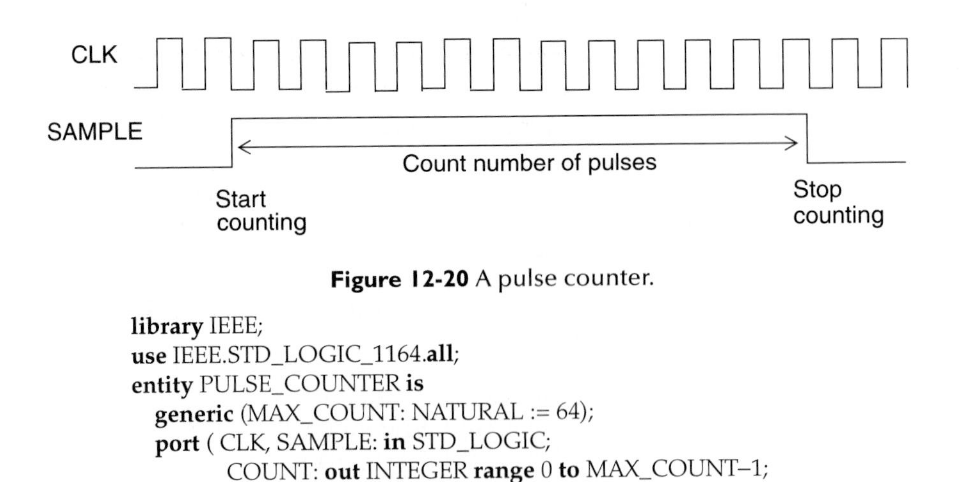

Figure 12-20 A pulse counter.

```
library IEEE;
use IEEE.STD_LOGIC_1164.all;
entity PULSE_COUNTER is
  generic (MAX_COUNT: NATURAL := 64);
  port ( CLK, SAMPLE: in STD_LOGIC;
         COUNT: out INTEGER range 0 to MAX_COUNT–1;
```

```
         OVERFLOW: out BOOLEAN);
end PULSE_COUNTER;

architecture BEHAVIOR of PULSE_COUNTER is
begin
  process (CLK, SAMPLE)
    variable LOCAL_CNTR: NATURAL;
    variable START_COUNTING: BOOLEAN;
  begin
    if SAMPLE'EVENT then
      if SAMPLE = '1' then
        -- Reset on rising edge.
        OVERFLOW <= FALSE;
        START_COUNTING := TRUE;
        LOCAL_CNTR := 0;
      elsif SAMPLE = '0' then
        START_COUNTING := FALSE;
      end if;
    elsif CLK'EVENT and CLK = '1' then
      -- Rising edge of clock.
      if START_COUNTING then
        LOCAL_CNTR := LOCAL_CNTR + 1;

        if LOCAL_CNTR > MAX_COUNT then
          LOCAL_CNTR := 0;
          OVERFLOW <= TRUE;
          START_COUNTING := FALSE;
        end if;
      end if;
    end if;
    COUNT <= LOCAL_CNTR;
  end process;
end;
```

After the SAMPLE has gone low, the counter value is retained until another rising edge occurs on SAMPLE.

Test Bench

Here is a test bench that exercises the above design. The clock is internally generated within the test bench but the SAMPLE waveform is read from a text file.

```
library IEEE;
use IEEE.STD_LOGIC_1164.all;
use STD.TEXTIO.all;
entity PULSE_TB is
```

```
  generic (MAX_CNT: POSITIVE := 20;
           OFF_PERIOD: TIME := 3 ns;
           ON_PERIOD: TIME := 2 ns);
end;

architecture TB of PULSE_TB is
  component PULSE_COUNTER
    generic (MAX_COUNT: NATURAL);
    port ( CLK, SAMPLE: in STD_LOGIC;
           COUNT: out INTEGER range 0 to MAX_COUNT–1;
           OVERFLOW: out BOOLEAN);
  end component;
  signal CLK, SAMPLE: STD_LOGIC;
  signal COUNT: INTEGER range 0 to MAX_CNT–1;
  signal OVERFLOW: BOOLEAN;
  signal STOP_CLK; BOOLEAN := FALSE;
begin
  PC: PULSE_COUNTER
        generic map (MAX_COUNT => MAX_CNT)
        port map (CLK, SAMPLE, COUNT, OVERFLOW);

  GEN_CLK: process
  begin
    CLK <= '0';
    wait for OFF_PERIOD;
    CLK <= '1';
    wait for ON_PERIOD;
    -- Stop clock if finished reading file:
    if STOP_CLK then
      wait;
    end if;
  end process;

  -- Read SAMPLE signal from text file that has entries of form:
  -- 0 ns     0
  -- 10 ns    1
  -- 25 ns    0
  -- 50 ns    1
  -- 100 ns   0

  process
    file SAM_WAVE: TEXT open READ_MODE
                        is "/home/bhasker/sample.wave";
    variable INBUF: LINE;
    variable AT_TIME: TIME;
    variable SAM_VALUE: BIT;
  begin
    while not ENDFILE (SAM_WAVE) loop
```

```
      READLINE (SAM_WAVE, INBUF);
      READ (INBUF, AT_TIME);                  -- Read time value.
      READ (INBUF, SAM_VALUE);                -- Line 4.
      wait for (AT_TIME – NOW);             -- Wait for specified time.
      -- Convert BIT value to STD_LOGIC type:
      SAMPLE = TO_STDULOGIC(SAM_VALUE);
      wait for 1 ns; -- Wait for SAMPLE to propagate through design.
      report "At time " & TIME'IMAGE(NOW) & " SAMPLE = " &
            STD_LOGIC'IMAGE(SAMPLE) & " COUNT = " &
            INTEGER'IMAGE(COUNT) &
            " OVERFLOW = " & BOOLEAN'IMAGE(OVERFLOW);
    end loop;
    STOP_CLK <= TRUE;
    wait;
  end process;
end;
```

The predefined READ procedure cannot read a STD_LOGIC value (see "Line 4"). Thus SAM_VALUE is read in as a BIT value and then it is converted to a STD_LOGIC type. The report statement is used to print out the values of the design. Alternately, WRITE procedure calls could also have been used. Here is a sample output produced following simulation.

```
** Note: At time 1 ns SAMPLE = '0' COUNT = 0 OVERFLOW = false
** Note: At time 11 ns SAMPLE = '1' COUNT = 0 OVERFLOW = false
** Note: At time 26 ns SAMPLE = '0' COUNT = 3 OVERFLOW = false
** Note: At time 51 ns SAMPLE = '1' COUNT = 0 OVERFLOW = false
** Note: At time 101 ns SAMPLE = '0' COUNT = 10 OVERFLOW = false
```

12.17 A Barrel Shifter

A barrel shifter shifts (actually rotates) the input data by the specified number of bits in a combinational manner. Here is a model for such a barrel shifter. Two architecture bodies are presented showing alternate ways of describing the same behavior and illustrating the power of VHDL.

```
library IEEE;
use IEEE.STD_LOGIC_1164.all;
entity BARREL_SHIFTER is
  generic (N: POSITIVE := 4);
  port ( DATA: in STD_LOGIC_VECTOR (N–1 downto 0);
```

```
            SELEKT: in INTEGER range 0 to N–1;
            BARR_OUT: out STD_LOGIC_VECTOR (N–1 downto 0));
end BARREL_SHIFTER;

architecture FOR_LOOP_STYLE of BARREL_SHIFTER is
begin
  process (SELEKT, DATA)
    variable VAR_BUF: STD_LOGIC_VECTOR(N–1 downto 0);
  begin
    VAR_BUF := DATA;

    for K in 1 to SELEKT loop
      VAR_BUF := VAR_BUF (N–2 downto 0) & VAR_BUF (N–1);
    end loop;

    BARR_OUT <= VAR_BUF;
  end process;
end FOR_LOOP_STYLE;

architecture CONCURRENT_STYLE of BARREL_SHIFTER is
begin
  BARR_OUT <=     DATA (N–1–SELEKT downto 0) &
                  DATA (N–1 downto N–SELEKT)
                  when SELEKT > 0
                  else DATA;
end CONCURRENT_STYLE;
```

12.18 Hierarchy in Design

A hierarchy in a design is managed by using component instantiation and component declaration statements. Given a large design, one could decompose a design into a number of smaller subdesigns. These components instantiate components at lower levels in the hierarchy. The description of components at any level could be in any design style, that is, dataflow, behavioral, structural, or mixed. For example, all components could initially be described as sequential processes, and as each component at each level is synthesized, it may be described using the structural style. Consider the hierarchical design shown in Figure 12-21.

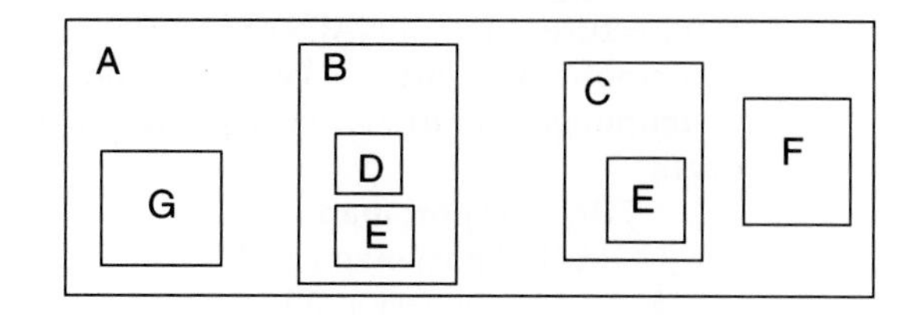

Figure 12-21 A hierarchical design.

Design A comprises four components, B, C, F, and G. Subdesign B comprises components D, E, and some glue behavioral logic. Subdesign C is composed of component E and some other behavioral logic. Component E is used by components B and C. Initially, there may be a separate entity declaration plus architecture body for each component, A, B, C, D, E, F, and G. As each component is synthesized to the gate level, different architecture bodies for each component are developed. A skeleton model for this large design example is shown next.

```
--Library and use clauses, if any.
entity A is port (. . .) end A;

architecture A_INTERNALS of A is
   component B port (. . .) end component;
   component C port (. . .) end component;
   component F port (. . .) end component;
   component G port (. . .) end component;
begin
   B_COMP1: B port map (. . .);
   C_COMP1: C port map (. . .);
   F_COMP1: F port map (. . .);
   G_COMP1: G port map (. . .);
   -- There might be other glue logic here, for example, high
   -- power drivers for the I/O for A, etc.
end A _INTERNALS;

entity F is port (. . .) end F;
architecture F_INTERNALS of F is
begin
   -- Behavior of F.
end F_INTERNALS;

entity G is port (. . .) end;
architecture G_INTERNALS of G is
begin
   -- Behavior of G.
end G_INTERNALS;
```

```
entity B is port (. . .) end B;
architecture B_INTERNALS of B is
   component D port (. . .) end component;
   component E port (. . .) end component;
begin
   D_COMP1: D port map (. . .);
   E_COMP1: E port map (. . .);
   -- Remaining behavior of B.
end B_INTERNALS;

entity C is port (. . .) end C;
architecture C_INTERNALS of C is
   component E port (. . .) end component;
begin
   E_COMP2: E port map (. . .);
   -- Other behavior of C.
end C_INTERNALS;

entity D is port (. . .) end D;
architecture D_INTERNALS of D is
begin
   -- Behavior of D.
end D_INTERNALS;

entity E is port (. . .) end E;
architecture E_INTERNALS of E is
begin
   -- Behavior of E.
end E_INTERNALS;
```

Each entity (entity declaration plus architecture body) could be placed in a separate text file and compiled into a working library. Having described the entire system behavior for design A, it could then be simulated and verified. There is no need to specify a configuration if all the entities are compiled into the working library, since the default bindings will be used (see Section 7.5).

Next, the synthesis of each component could proceed either top-down, that is, translate behavior of design A to structure, then for B, C, D, E, F, G, or it can proceed bottom-up, that is, synthesize subdesign E and D before B or C. Say component E is synthesized into a structure, either using available synthesis tools or by a manual design process; this structure could be saved in a new architecture body that is still associated with the same entity declaration E. For example,

```
architecture E_2 of E is
begin
```

```
    -- Structure of E.
end E_2;
```

To simulate this new architecture, a configuration declaration would have to be written that would bind entity E to architecture E_2. For example,

```
configuration E_CON of E is
  for E_2
  end for;
end E_CON;
```

Configuration E_CON would now be verified on a stand-alone basis. This would verify the entity E using architecture E_2. To simulate the entire design A again, references to component E in design B and C must bind its instantiations to the new architecture for E. This could be done using a configuration specification that is added to the architecture body for both B and C, such as:

```
for comp-label : E use configuration WORK.E_CON;
```

If the architecture bodies for B and C cannot be changed, a configuration declaration can be written for design A, as shown next.

```
configuration A_CON of A is
  for B
    for E_COMP1: E
      use configuration WORK.E_CON;
    end for;
  end for;
  for C
    for E_COMP2: E
      use configuration WORK.E_CON;
    end for;
  end for;
  -- For unbound components, use default rules (see Section 7.5).
end A_CON;
```

The basic idea is that every time a component is synthesized into its structure, a new architecture body is created that is associated with the same entity declaration. A configuration declaration or a configuration specification can be written to bind the new architecture with the component one level above in the hierarchy.

It is not necessary to keep the subdesigns of an entire design in a single design library. However, all architecture bodies of a component must reside in the same library. This is because they must reside with their corresponding entity declarations. If components reside in different libraries, configuration specifications have to be written (the default rules work for only components in one library, that is, the working library), and these are made visible to their next higher-level component by using context clauses. For example, if component E were to reside in library JOHNS_LIB, the configuration specification appearing in the architecture for subdesign C may be:

```
library JOHNS_LIB;
architecture C_2 of C is
  for E_COMP2: E use configuration JOHNS_LIB.E_CON;
begin
  . . .
end C_2;
```

❑

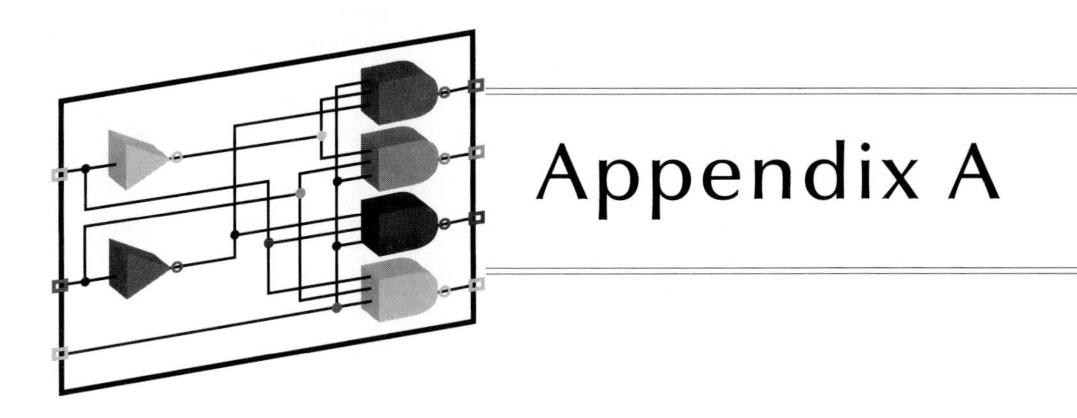

Appendix A

Predefined Environment

This appendix describes the predefined environment of the language. The set of reserved words for the language is listed in the first section. The next two sections give the source code listing for the predefined packages STANDARD and TEXTIO. The predefined attributes have already been described in Chapter 10. All information in this appendix has been reprinted from the IEEE Std 1076-1993 VHDL Language Reference Manual.[1]

A.1 Reserved Words

The following identifiers are reserved words in the language (also called keywords), and therefore cannot be used as basic identifiers in a VHDL description.

1. Reprinted here from the IEEE Std 1076-1993 VHDL LRM by permission from IEEE.

abs **access** **after** **alias**
all **and** **architecture** **array**
assert **attribute**

begin **block** **body** **buffer**
bus

case **component** **configuration** **constant**

disconnect **downto**

else **elsif** **end** **entity**
exit

file **for** **function**

generate **generic** **group** **guarded**

if **impure** **in** **inertial**
inout **is**

label **library** **linkage** **literal**
loop

map **mod**

nand **new** **next** **nor**
not **null**

of **on** **open** **or**
others **out**

package **port** **postponed** **procedure**
process **pure**

range **record** **register** **reject**
rem **report** **return** **rol**
ror

select **severity** **signal** **shared**
sla **sll** **sra** **srl**
subtype

then **to** **transport** **type**

unaffected **units** **until** **use**

variable

wait **when** **while** **with**

xnor **xor**

A.2 Package STANDARD

The package STANDARD is a predefined package that contains the definitions for the predefined types and functions of the language. A source code listing of this package follows.

```
package STANDARD is
  -- Predefined enumeration types:
  type BOOLEAN is (FALSE, TRUE);
  type BIT is ('0', '1');
  type CHARACTER is (
        NUL,  SOH,  STX,  ETX,  EOT,  ENQ,  ACK,
        BEL,  BS,   HT,   LF,   VT,   FF,   CR,
        SO,   SI,   DLE,  DC1,  DC2,  DC3,  DC4,
        NAK,  SYN,  ETB,  CAN,  EM,   SUB,  ESC,
        FSP,  GSP,  RSP,  USP,

        ' ',  '!',  '"',  '#',  '$',  '%',  '&',
        ''',  '(',  ')',  '*',  '+',  ',',  '-',
        '.',  '/',  '0',  '1',  '2',  '3',  '4',
        '5',  '6',  '7',  '8',  '9',  ':',  ';',
        '<',  '=',  '>',  '?',

        '@',  'A',  'B',  'C',  'D',  'E',  'F',
        'G',  'H',  'I',  'J',  'K',  'L',  'M',
        'N',  'O',  'P',  'Q',  'R',  'S',  'T',
        'U',  'V',  'W',  'X',  'Y',  'Z',  '[',
        '\',  ']',  '^',  '_',

        '`',  'a',  'b',  'c',  'd',  'e',  'f',
        'g',  'h',  'i',  'j',  'k',  'l',  'm',
        'n',  'o',  'p',  'q',  'r',  's',  't',
        'u',  'v',  'w',  'x',  'y',  'z',  '{',
        '|',  '}',
```

```
            -- Plus other characters from
            -- the ISO² 8859-1:1987(E) standard.
            );

type SEVERITY_LEVEL is (NOTE, WARNING,
            ERROR, FAILURE);

-- Predefined numeric types:
type INTEGER is range implementation_defined;
type REAL is range implementation_defined;

-- Predefined physical type TIME:
type TIME is range implementation_defined
  units
    fs;                              -- femtosecond
    ps   = 1000 fs;                  -- picosecond
    ns   = 1000 ps;                  -- nanosecond
    us   = 1000 ns;                  -- microsecond
    ms   = 1000 us;                  -- microsecond
    sec  = 1000 ms;                  -- seconds
    min  = 60 secs;                  -- minutes
    hr   = 60 min;                   -- hours
  end units;

-- Predefined physical subtype:
subtype DELAY_LENGTH is TIME range 0 fs to TIME'HIGH;

[illegible]nction that returns the current simulation time:
  [illegible]fu[illegible]LAY_LENGTH;

-- Predefined numeric subtypes:
subtype NATURAL is INTEGER range 0 to [illegible]TEGER'HIGH;
subtype POSITIVE is INTEGER range 1 to I[illegible]TEGER'HIGH;

-- Predefined array types:
type STRING is array (POSITIVE range <>) [illegible] CHARACTER;
type BIT_VECTOR is array (NATURAL rang[illegible]>) of BIT;

-- Predefined types for file operations:
type FILE_OPEN_KIND
            is (READ_MODE, WRITE_MO[illegible], APPEND_MODE);
type FILE_OPEN_STATUS is (OPEN_OK,
            STATUS_ERROR, NAME_ERR[illegible], MODE_ERROR);

-- Attribute declaration:
attribute FOREIGN: STRING;
end STANDARD;
```

2. I[illegible]s Organization.

Package TEXTIO contains dec[illegible]rations of types and subprograms that support formatted I/O operations on text files. A source code listing of the package follows.

```
package TEXTIO is
   -- Type definitions for text I/O:
   type LINE is access STRING;          -- A line is a pointer to a
                                        -- STRING value.
   type TEXT is file of STRING;         -- A file of variable-length
                                        -- ASCII records.
   type SIDE is (RIGHT, LEFT);          -- For justifying output data
                                        -- within fields.
   subtype WIDTH is NATURAL;            -- For specifying widths of
                                        -- output fields.

   -- Standard text files:
   file INPUT: TEXT open READ_OPEN is "STD_INPUT";
   file OUTPUT: TEXT open WRITE_OPEN is "STD_OUTPUT";

   -- Input routines for standard types:
   procedure RE[illegible]LINE (file [illegible] TEXT; L: [illegible]

   [illegible] VALUE: [illegible]
                     GOOD: out BOOLEAN);
   procedure READ (L: inout LINE; VALUE: out BI[illegible]

   procedure READ (L: inout LINE; VALUE: out BI[illegible]OR;
                     GOOD: out BOOLEAN);
   procedure READ (L: inout LINE; VALUE: out BI[illegible]OR);

   procedure READ (L: inout LINE; VALUE: out BO[illegible]N;
                     GOOD: out BOOLEAN);
   procedure READ (L: inout LINE; VALUE: out BO[illegible]N);

   procedure READ (L: inout LINE; VALUE: out CH[illegible]ER;
                     GOOD: out BOOLEAN);
   procedure READ (L: inout LINE; VALUE: out C[illegible]ER);

   procedure READ (L: inout LINE; VALUE: out IN[illegible]
                     GOOD: out BOOLEAN);
   procedure READ (L: inout LINE; VALUE: out IN[illegible]

   procedure READ (L: inout LINE; VALUE: out RE[illegible]
                     GOOD: out BOOLEAN);
```

```
procedure READ (L: inout LINE; VALUE: out REAL);

procedure READ (L: inout LINE; VALUE: out STRING;
                GOOD: out BOOLEAN);
procedure READ (L: inout LINE; VALUE: out STRING);

procedure READ (L: inout LINE; VALUE: out TIME;
                GOOD: out BOOLEAN);
procedure READ (L: inout LINE; VALUE: out TIME);

-- Output routines for standard types:
procedure WRITELINE (file F: TEXT; L: in LINE);

procedure WRITE (L: inout LINE; VALUE: in BIT;
            JUSTIFIED: in SIDE := RIGHT; FIELD: in WIDTH := 0);
procedure WRITE (L: inout LINE; VALUE: in BIT_VECTOR;
            JUSTIFIED: in SIDE := RIGHT; FIELD: in WIDTH := 0);
procedure WRITE (L: inout LINE; VALUE: in BOOLEAN;
            JUSTIFIED: in SIDE := RIGHT; FIELD: in WIDTH := 0);
procedure WRITE (L: inout LINE; VALUE: in CHARACTER;
            JUSTIFIED: in SIDE := RIGHT; FIELD: in WIDTH := 0);
procedure WRITE (L: inout LINE; VALUE: in INTEGER;
            JUSTIFIED: in SIDE := RIGHT; FIELD: in WIDTH := 0);
procedure WRITE (L: inout LINE; VALUE: in REAL;
            JUSTIFIED: in SIDE := RIGHT; FIELD: in WIDTH := 0;
            DIGITS: in NATURAL := 0);
procedure WRITE (L: inout LINE; VALUE: in STRING;
            JUSTIFIED: in SIDE := RIGHT; FIELD: in WIDTH := 0);
procedure WRITE (L: inout LINE; VALUE: in TIME;
            JUSTIFIED: in SIDE := RIGHT; FIELD: in WIDTH := 0;
            UNIT: in TIME := ns);

-- File position predicate:
-- function ENDFILE (file F: TEXT) return BOOLEAN;
end TEXTIO;
```

❑

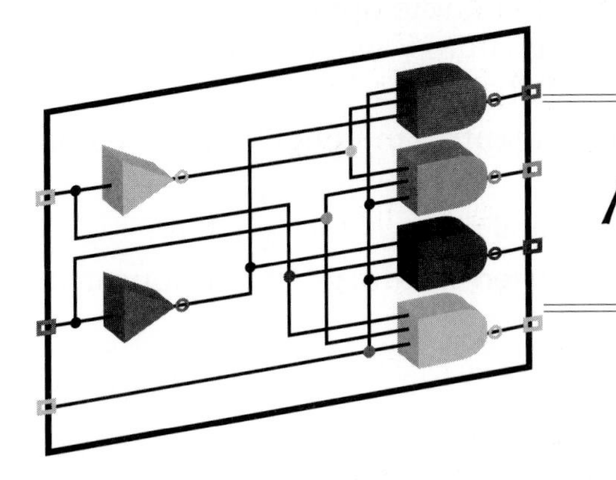

Appendix B

Syntax Reference

This appendix presents the complete syntax of the VHDL language.[1]

B.1 Conventions

The following conventions are used in describing this syntax.

1. The syntax rules are organized in an alphabetical order by their left-hand nonterminal name.
2. Reserved words are written in **boldface**.
3. A name in *italics* prefixed to a nonterminal name represents the semantic meaning associated with that nonterminal name.
4. The vertical bar symbol (|) separates alternative items unless it appears immediately after an opening brace, in which case it stands for itself.

1. Reprinted here from the IEEE Std 1076-1993 VHDL LRM by permission from IEEE.

5. Square brackets ([. . .]) denote optional items, except in the syntax for "signature", where the outermost brackets are required characters.
6. Curly braces ({ . . . }) identify an item that is repeated zero or more times.
7. The starting nonterminal name is "design_file".
8. The terminal names used in this grammar appear in upper case.

B.2 The Syntax

abstract_literal ::=
  decimal_literal
  | based_literal

access_type_definition ::= **access** subtype_indication

actual_designator ::=
  expression
  | *signal_*name
  | *variable_*name
  | **open**

actual_parameter_part := *parameter_*association_list

actual_part ::=
  actual_designator
  | *function_*name (actual_designator)
  | type_mark (actual_designator)

adding_operator ::= + | – | &

aggregate ::= (element_association { , element_association })

alias_declaration ::=
  alias alias_designator [: subtype_indication] **is** name [signature] ;

alias_designator ::= identifier | character_literal | operator_symbol

allocator ::= **new** subtype_indication | **new** qualified_expression

architecture_body ::=
  architecture identifier **of** *entity_*name **is**
    architecture_declarative_part
  begin
    architecture_statement_part
  end [*architecture_*simple_name] ;

architecture_declarative_part ::= { block_declarative_item }

architecture_statement_part ::= { concurrent_statement }

array_type_definition ::=
    unconstrained_array_definition | constrained_array_definition

assertion ::=
    assert condition
        [**report** expression]
        [**severity** expression]

assertion_statement ::= [label :] assertion ;

association_element ::= [formal_part =>] actual_part

association_list ::= association_element { , association_element }

attribute_declaration ::= **attribute** identifier : type_mark ;

attribute_designator ::= *attribute*_simple_name

attribute_name ::= prefix [signature] ' attribute_designator [(expression)]

attribute_specification ::=
    attribute attribute_designator **of** entity_specification **is** expression ;

base ::= integer

base_specifier ::= B | O | X

base_unit_declaration ::= identifier ;

based_integer ::= extended_digit { [UNDERLINE] extended_digit }

based_literal ::= base # based_integer [. based_integer] # [exponent]

basic_character ::= basic_graphic_character | FORMAT_EFFECTOR

basic_graphic_character :=
    UPPER_CASE_LETTER
    | DIGIT
    | SPECIAL_CHARACTER
    | SPACE_CHARACTER

basic_identifier ::= letter { [UNDERLINE] letter_or_digit }

binding_indication ::=
    [**use** entity_aspect] [generic_map_aspect] [port_map_aspect]

bit_string_literal ::= base_specifier " [bit_value] "

bit_value ::= extended_digit { [UNDERLINE] extended_digit }

block_configuration ::=
    for block_specification
        { use_clause }
        { configuration_item }
    end for;

block_declarative_item ::=
  subprogram_declaration
  | subprogram_body
  | type_declaration
  | subtype_declaration
  | constant_declaration
  | signal_declaration
  | *shared*_variable_declaration
  | file_declaration
  | alias_declaration
  | component_declaration
  | attribute_declaration
  | attribute_specification
  | configuration_specification
  | disconnection_specification
  | use_clause
  | group_template_declaration
  | group_declaration

block_declarative_part ::= { block_declarative_item }

block_header ::=
  [generic_clause [generic_map_aspect ;]]
  [port_clause [port_map_aspect ;]]

block_specification ::=
  *architecture*_name
  | *block_statement*_label
  | *generate_statement*_label [(index_specification)]

block_statement ::=
  *block*_label :
    block [(*guard*_expression)] [**is**]
      block_header
      block_declarative_part
    begin
      block_statement_part
    end block [*block*_label] ;

block_statement_part ::= { concurrent_statement }

case_statement ::=
  [*case*_label :]
  case expression **is**
    case_statement_alternative
    { case_statement_alternative }
  end case [*case*_label] ;

case_statement_alternative ::= **when** choices => sequence_of_statements

character_literal ::= ' graphic_character '

choice ::=
    simple_expression
    | discrete_range
    | *element*_simple_name
    | **others**

choices ::= choice { | choice }

component_configuration ::=
    for component_specification
        [binding_indication ;]
        [block_configuration]
    end for;

component_declaration ::=
    component identifier [**is**]
        [*local*_generic_clause]
        [*local*_port_clause]
    end component [*component*_simple_name] ;

component_instantiation_statement ::=
    *instantiation*_label :
        instantiated_unit [generic_map_aspect] [port_map_aspect] ;

component_specification ::= instantiation_list : *component*_name

composite_type_definition ::= array_type_definition | record_type_definition

concurrent_assertion_statement ::= [label :] [**postponed**] assertion ;

concurrent_procedure_call_statement ::=
    [label :] [**postponed**] procedure_call ;

concurrent_signal_assignment_statement ::=
    [label :] [**postponed**] conditional_signal_assignment
    | [label :] [**postponed**] selected_signal_assignment

concurrent_statement ::=
    block_statement
    | process_statement
    | concurrent_procedure_call_statement
    | concurrent_assertion_statement
    | concurrent_signal_assignment_statement
    | component_instantiation_statement
    | generate_statement

condition ::= *boolean*_expression

condition_clause ::= **until** condition

conditional_signal_assignment ::= target <= options conditional_waveforms ;

conditional_waveforms ::=
    { waveform **when** condition **else** } waveform [**when** condition]

configuration_declaration ::=
  configuration identifier **of** *entity*_name **is**
    configuration_declarative_part
    block_configuration
  end configuration [*configuration*_simple_name] ;

configuration_declarative_item ::= use_clause | attribute_specification

configuration_declarative_part ::= { configuration_declarative_item }

configuration_item ::= block_configuration | component_configuration

configuration_specification ::=
  for component_specification binding_indication ;

constant_declaration ::=
  constant identifier_list : subtype_indication [:= expression] ;

constrained_array_definition ::=
  array index_constraint **of** *element*_subtype_indication

constraint ::= range_constraint | index_constraint

context_clause ::= { context_item }

context_item ::= library_clause | use_clause

decimal_literal ::= integer [. integer] [exponent]

declaration ::=
  type_declaration
  | subtype_declaration
  | object_declaration
  | interface_declaration
  | alias_declaration
  | attribute_declaration
  | component_declaration
  | group_template_declaration
  | group_declaration
  | entity_declaration
  | configuration_declaration
  | subprogram_declaration
  | package_declaration

delay_mechanism ::=
  transport
  | [**reject** *time*_expression] **inertial**

design_file ::= design_unit { design_unit }

design_unit ::= context_clause library_unit

designator ::= identifier | operator_symbol

direction ::= **to** | **downto**

disconnection_specification ::=
disconnect guarded_signal_specification **after** *time*_expression ;

discrete_range ::= *discrete*_subtype_indication | range

element_association ::= [choices =>] expression

element_declaration ::= identifier_list : element_subtype_definition ;

element_subtype_definition ::= subtype_indication

entity_aspect ::=
entity *entity*_name [(*architecture*_identifier)]
| **configuration** *configuration*_name
| **open**

entity_class ::=
entity | **architecture** | **configuration** | **procedure**
| **function** | **package** | **type** | **subtype** | **constant** | **signal**
| **variable** | **component** | **label** | **literal** | **units** | **group** | **file**

entity_class_entry ::= entity_class [<>]

entity_class_entry_list ::= entity_class_entry { , entity_class_entry }

entity_declaration ::=
entity identifier **is**
entity_header
entity_declarative_part
[**begin**
entity_statement_part]
end entity [*entity*_simple_name] ;

entity_declarative_item ::=
subprogram_declaration
| subprogram_body
| type_declaration
| subtype_declaration
| constant_declaration
| signal_declaration
| *shared*_variable_declaration
| file_declaration
| alias_declaration
| attribute_declaration
| attribute_specification
| disconnection_specification
| use_clause
| group_template_declaration
| group_declaration

entity_declarative_part ::= { entity_declarative_item }

entity_designator ::= entity_tag [signature]

entity_header ::= [*formal*_generic_clause] [*formal*_port_clause]

entity_name_list ::=
  entity_designator { , entity_designator }
  | **others**
  | **all**

entity_specification ::= entity_name_list : entity_class

entity_statement ::=
  concurrent_assertion_statement
  | *passive*_concurrent_procedure_call_statement
  | *passive*_process_statement

entity_statement_part ::= { entity_statement }

entity_tag ::= simple_name | character_literal | operator_symbol

enumeration_literal ::= identifier | character_literal

enumeration_type_definition ::= (enumeration_literal { , enumeration_literal })

exit_statement ::= [label :] **exit** [*loop*_label] [**when** condition] ;

exponent ::= E [+] integer | E – integer

expression ::=
  relation { **and** relation }
  | relation { **or** relation }
  | relation { **xor** relation }
  | relation [**nand** relation]
  | relation [**nor** relation]
  | relation [**xnor** relation]

extended_digit ::= DIGIT | letter

extended_identifier ::= \ graphic_character { graphic_character } \

factor ::=
  primary [** primary]
  | **abs** primary
  | **not** primary

file_declaration ::=
  file identifier_list : subtype_indication [file_open_information] ;

file_logical_name ::= *string*_expression

file_open_information ::=
  [**open** *file_open_kind*_expression] **is** file_logical_name

file_type_definition ::= **file of** type_mark

floating_type_definition ::= range_constraint

formal_designator ::= *generic*_name | *port*_name | *parameter*_name

formal_parameter_list ::= *parameter*_interface_list

formal_part ::=
    formal_designator
    | *function*_name (formal_designator)
    | type_mark (formal_designator)

full_type_declaration ::= **type** identifier **is** type_definition ;

function_call ::= *function*_name [(actual_parameter_part)]

generate_statement ::=
    *generate*_label :
        generation_scheme **generate**
            [{ block_declarative_item }
        begin]
            { concurrent_statement }
        end generate [*generate*_label] ;

generation_scheme ::=
    for *generate*_parameter_specification
    | **if** condition

generic_clause ::= **generic** (generic_list) ;

generic_list ::= *generic*_interface_list

generic_map_aspect ::= **generic map** (*generic*_association_list)

graphic_character ::=
    basic_graphic_character
    | LOWER_CASE_LETTER
    | OTHER_SPECIAL_CHARACTER

group_constituent ::= name | character_literal

group_constituent_list ::= group_constituent { , group_constituent }

group_template_declaration ::= **group** identifier **is** (entity_class_entry_list) ;

group_declaration ::=
    group identifier : *group_template*_name (group_constituent_list) ;

guarded_signal_specification ::= *guarded*_signal_list : type_mark

identifier ::= basic_identifier | extended_identifier

identifier_list ::= identifier { , identifier }

if_statement ::=
    [*if*_label :]
    if condition **then**
        sequence_of_statements
    { **elsif** condition **then**
        sequence_of_statements }
    [**else**
        sequence_of_statements]
    end if [*if*_label] ;

incomplete_type_declaration ::= **type** identifier ;

index_constraint ::= (discrete_range { , discrete_range })

index_specification ::= discrete_range | *static*_expression

index_subtype_definition ::= type_mark **range** <>

indexed_name ::= prefix (expression { , expression })

instantiated_unit ::=
    [**component**] *component*_name
    | **entity** *entity*_name [(*architecture*_identifier)]
    | **configuration** *configuration*_name

instantiation_list ::=
    *instantiation*_label { , *instantiation*_label }
    | **others**
    | **all**

integer ::= DIGIT { [UNDERLINE] DIGIT }

integer_type_definition ::= range_constraint

interface_constant_declaration ::=
    [**constant**] identifier_list : [**in**] subtype_indication
    [:= *static*_expression]

interface_declaration ::=
    interface_constant_declaration
    | interface_signal_declaration
    | interface_variable_declaration
    | interface_file_declaration

interface_element ::= interface_declaration

interface_file_declaration ::= **file** identifier_list : subtype_indication

interface_list ::= interface_element { ; interface_element }

interface_signal_declaration ::=
    [**signal**] identifier_list : [mode] subtype_indication [**bus**]
    [:= *static*_expression]

interface_variable_declaration ::=
    [**variable**] identifier_list : [mode] subtype_indication
    [:= *static*_expression]

iteration_scheme ::=
    while condition
    | **for** *loop*_parameter_specification

label ::= identifier

letter ::= UPPER_CASE_LETTER | LOWER_CASE_LETTER

letter_or_digit ::= letter | digit

library_clause ::= **library** logical_name_list ;

library_unit ::= primary_unit | secondary_unit

literal ::=
    numeric_literal
    | enumeration_literal
    | string_literal
    | bit_string_literal
    | **null**

logical_name ::= identifier

logical_name_list ::= logical_name { , logical_name }

logical_operator ::= **and** | **or** | **nand** | **nor** | **xor** | **xnor**

loop_statement ::=
    [*loop*_label :]
      [iteration_scheme] **loop**
        sequence_of_statements
      end loop [*loop*_label] ;

miscellaneous_operator ::= ** | **abs** | **not**

mode ::= **in** | **out** | **inout** | **buffer** | **linkage**

multiplying_operator ::= * | / | **mod** | **rem**

name ::=
    simple_name
    | operator_symbol
    | selected_name
    | indexed_name
    | slice_name
    | attribute_name

next_statement ::= [label :] **next** [*loop*_label] [**when** condition] ;

null_statement ::= [label :] **null** ;

numeric_literal ::= abstract_literal | physical_literal

object_declaration ::=
    constant_declaration
    | signal_declaration
    | variable_declaration
    | file_declaration

operator_symbol ::= string_literal

options ::= [**guarded**] [delay_mechanism]

package_body ::=
    package body *package*_simple_name **is**

package_body_declarative_part
end [**package body**] [*package*_simple_name] ;

package_body_declarative_item ::=
subprogram_declaration
| subprogram_body
| type_declaration
| subtype_declaration
| constant_declaration
| *shared*_variable_declaration
| file_declaration
| alias_declaration
| use_clause
| group_template_declaration
| group_declaration

package_body_declarative_part ::= { package_body_declarative_item }

package_declaration ::=
package identifier **is**
package_declarative_part
end [**package**] [*package*_simple_name] ;

package_declarative_item ::=
subprogram_declaration
| type_declaration
| subtype_declaration
| constant_declaration
| signal_declaration
| *shared*_variable_declaration
| file_declaration
| alias_declaration
| component_declaration
| attribute_declaration
| attribute_specification
| disconnection_specification
| use_clause
| group_template_declaration
| group_declaration

package_declarative_part ::= { package_declarative_item }

parameter_specification ::= identifier **in** discrete_range

physical_literal ::= [abstract_literal] *unit*_name

physical_type_definition ::=
range_constraint
units
base_unit_declaration

{ secondary_unit_declaration }
end units [*physical_type*_simple_name]

port_clause ::= **port** (port_list) ;

port_list ::= *port*_interface_list

port_map_aspect ::= **port map** (*port*_association_list)

prefix ::= name | function_call

primary ::=
name
| literal
| aggregate
| function_call
| qualified_expression
| type_conversion
| allocator
| (expression)

primary_unit ::=
entity_declaration
| configuration_declaration
| package_declaration

procedure_call ::= *procedure*_name [(actual_parameter_part)]

procedure_call_statement ::= [label :] procedure_call ;

process_declarative_item ::=
subprogram_declaration
| subprogram_body
| type_declaration
| subtype_declaration
| constant_declaration
| variable_declaration
| file_declaration
| alias_declaration
| attribute_declaration
| attribute_specification
| use_clause
| group_template_declaration
| group_declaration

process_declarative_part ::= { process_declarative_item }

process_statement ::=
[*process*_label :]
[**postponed**] **process** [(sensitivity_list)] [**is**]
process_declarative_part
begin

process_statement_part
end [**postponed**] **process** [*process*_label] ;

process_statement_part ::= { sequential_statement }

qualified_expression ::=
type_mark ' (expression)
| type_mark ' aggregate

range ::=
*range*_attribute_name
| simple_expression direction simple_expression

range_constraint ::= **range** range

record_type_definition ::=
record
element_declaration
{ element_declaration }
end record [*record_type*_simple_name]

relation ::= shift_expression [relational_operator shift_expression]

relational_operator ::= = | /= | < | <= | > | >=

report_statement ::= [label :] **report** expression [**severity** expression] ;

return_statement ::= [label :] **return** [expression] ;

scalar_type_definition ::=
enumeration_type_definition
| integer_type_definition
| floating_type_definition
| physical_type_definition

secondary_unit ::= architecture_body | package_body

secondary_unit_declaration ::= identifier = physical_literal ;

selected_name ::= prefix . suffix

selected_signal_assignment ::=
with expression **select**
target <= options selected_waveforms ;

selected_waveforms ::= { waveform **when** choices , } waveform **when** choices

sensitivity_clause ::= **on** sensitivity_list

sensitivity_list ::= *signal*_name { , *signal*_name }

sequence_of_statements ::= { sequential_statement }

sequential_statement ::=
wait_statement
| assertion_statement
| report_statement

| signal_assignment_statement
| variable_assignment_statement
| procedure_call_statement
| if_statement
| case_statement
| loop_statement
| next_statement
| exit_statement
| return_statement
| null_statement

shift_expression ::= simple_expression [shift_operator simple_expression]

shift_operator ::= **sll** | **srl** | **sla** | **sra** | **rol** | **ror**

sign ::= + | –

signal_assignment_statement ::=
[label :] target <= [delay_mechanism] waveform ;

signal_declaration ::=
signal identifier_list : subtype_indication [signal_kind] [:= expression] ;

signal_kind ::= **register** | **bus**

signal_list ::=
*signal*_name { , *signal*_name }
| **others**
| **all**

signature[2] ::= [[type_mark { , type_mark }] [**return** type_mark]]

simple_expression ::= [sign] term { adding_operator term }

simple_name ::= identifier

slice_name ::= prefix (discrete_range)

string_literal ::= " { graphic_character } "

subprogram_body ::=
subprogram_specification **is**
subprogram_declarative_part
begin
subprogram_statement_part
end [subprogram_kind] [designator] ;

subprogram_declaration ::= subprogram_specification ;

subprogram_declarative_item ::=
subprogram_declaration

2. The outermost square brackets are part of the syntax and do not represent an optional item.

| subprogram_body
| type_declaration
| subtype_declaration
| constant_declaration
| variable_declaration
| file_declaration
| alias_declaration
| attribute_declaration
| attribute_specification
| use_clause
| group_template_declaration
| group_declaration

subprogram_declarative_part ::= { subprogram_declarative_item }

subprogram_kind ::= **procedure** | **function**

subprogram_specification ::=
procedure designator [(formal_parameter_list)]
| [**pure** | **impure**] **function** designator [(formal_parameter_list)]
return type_mark

subprogram_statement_part ::= { sequential_statement }

subtype_declaration ::= **subtype** identifier **is** subtype_indication ;

subtype_indication ::= [*resolution_function_*name] type_mark [constraint]

suffix ::=
simple_name
| character_literal
| operator_symbol
| **all**

target ::= name | aggregate

term ::= factor { multiplying_operator factor }

timeout_clause := **for** *time_*expression

type_conversion ::= type_mark (expression)

type_declaration ::= full_type_declaration | incomplete_type_declaration

type_definition ::=
scalar_type_definition
| composite_type_definition
| access_type_definition
| file_type_definition

type_mark ::= *type_*name | *subtype_*name

unconstrained_array_definition ::=
array (index_subtype_definition { , index_subtype_definition })
of *element_*subtype_indication

use_clause ::= **use** selected_name { , selected_name } ;

variable_assignment_statement ::= [label :] target := expression ;

variable_declaration ::=
  [**shared**] **variable** identifier_list : subtype_indication [:= expression] ;

wait_statement ::=
  [label :] **wait** [sensitivity_clause]
    [condition_clause]
    [timeout_clause] ;

waveform ::=
  waveform_element { , waveform_element }
  | **unaffected**

waveform_element ::=
  *value_*expression [**after** *time_*expression]
  | **null** [**after** *time_*expression]

❑

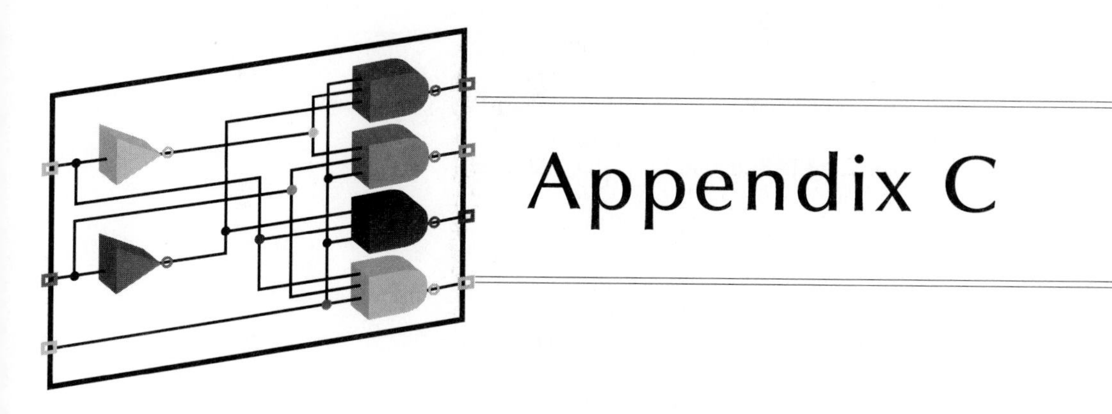

Appendix C

A Package Example

This appendix describes the complete ATT_MVL package that has been referred to in previous editions of this book. The package is provided more as an illustration of what goes into a package rather than an attempt to present a single comprehensive package.

C.1 The Package ATT_MVL

The ATT_MVL contains the definition of a four-value type called MVL and its associated overloaded logical operators. Here is the VHDL source code listing for this package.

```
package ATT_MVL is
   type MVL is ('U', '0', '1', 'Z');
   type MVL_VECTOR is array (NATURAL range <>) of MVL;

   type MVL_1D_TABLE is array (MVL) of MVL;
   type MVL_2D_TABLE is array (MVL, MVL) of MVL;

   -- Truth tables for logical operators:
   constant TABLE_AND: MVL_2D_TABLE :=
```

```
   -- U      0      1      Z
   (( 'U',   '0',   'U',   'U'),    -- U
    ( '0',   '0',   '0',   '0'),    -- 0
    ( 'U',   '0',   '1',   'U'),    -- 1
    ( 'U',   '0',   'U',   'U'));   -- Z
constant TABLE_OR: MVL_2D_TABLE :=
   -- U      0      1      Z
   (( 'U',   'U',   '1',   'U'),    -- U
    ( 'U',   '0',   '1',   'U'),    -- 0
    ( '1',   '1',   '1',   '1'),    -- 1
    ( 'U',   'U',   '1',   'U'));   -- Z
constant TABLE_NAND: MVL_2D_TABLE :=
   -- U      0      1      Z
   (( 'U',   '1',   'U',   'U'),    -- U
    ( '1',   '1',   '1',   '1'),    -- 0
    ( 'U',   '1',   '0',   'U'),    -- 1
    ( 'U',   '1',   'U',   'U'));   -- Z
constant TABLE_NOR: MVL_2D_TABLE :=
   -- U      0      1      Z
   (( 'U',   'U',   '0',   'U'),    -- U
    ( 'U',   '1',   '0',   'U'),    -- 0
    ( '0',   '0',   '0',   '0'),    -- 1
    ( 'U',   'U',   '0',   'U'));   -- Z
constant TABLE_XOR: MVL_2D_TABLE :=
   -- U      0      1      Z
   (( 'U',   'U',   'U',   'U'),    -- U
    ( 'U',   '0',   '1',   'U'),    -- 0
    ( 'U',   '1',   '0',   'U'),    -- 1
    ( 'U',   'U',   'U',   'U'));   -- Z
constant TABLE_NOT: MVL_1D_TABLE :=
   -- U      0      1      Z
   ( 'U',    '1',   '0',   'U');
constant TABLE_BUF: MVL_1D_TABLE :=
   -- U      0      1      Z
   ( 'U',    '0',   '1',   'U');

-- Truth tables for resolution functions:
constant TABLE_TAND: MVL_2D_TABLE :=
   -- U      0      1      Z
   (( 'U',   '0',   'U',   'U'),    -- U
    ( '0',   '0',   '0',   '0'),    -- 0
    ( 'U',   '0',   '1',   '1'),    -- 1
    ( 'U',   '0',   '1',   'Z'));   -- Z

constant TABLE_TOR: MVL_2D_TABLE :=
   --U       0      1      Z
   (( 'U',   'U',   '1',   'U'),    -- U
```

```
    ( 'U',   '0',   '1',   '0'),    -- 0
    ( '1',   '1',   '1',   '1'),    -- 1
    ( 'U',   '0',   '1',   'Z'));   -- Z

  -- Overloaded logical operator declarations on MVL type:
  function "and"   (L, R: MVL) return MVL;
  function "or"    (L, R: MVL) return MVL;
  function "nand"  (L, R: MVL) return MVL;
  function "nor"   (L, R: MVL) return MVL;
  function "xor"   (L, R: MVL) return MVL;
  function "not"   (L: MVL) return MVL;

  -- Overloaded logical operator declarations on MVL_VECTOR type:
  function "and"   (L, R: MVL_VECTOR) return MVL_VECTOR;
  function "or"    (L, R: MVL_VECTOR) return MVL_VECTOR;
  function "nand"  (L, R: MVL_VECTOR) return MVL_VECTOR;
  function "nor"   (L, R: MVL_VECTOR) return MVL_VECTOR;
  function "xor"   (L, R: MVL_VECTOR) return MVL_VECTOR;
  function "not"   (L: MVL_VECTOR) return MVL_VECTOR;

  -- Common utilities:
  function MAX     (T1, T2: TIME) return TIME;
  function INT2MVL (OPD, NO_BITS: INTEGER)
                        return MVL_VECTOR;
  function MVL2INT (OPD: MVL_VECTOR) return INTEGER;
  procedure P_1 (signal DRIVE_SIGNAL: inout MVL;
                 SIGNAL_VALUE: in MVL;
                 DRIVE_WIDTH, DRIVE_DELAY: in TIME);

  -- Overloaded arithmetic operators on MVL_VECTOR:
  function "+"   (L, R: MVL_VECTOR) return MVL_VECTOR;
  function "*"   (L, R: MVL_VECTOR) return MVL_VECTOR;

  -- Clock functions:
  function ES_RISING     (signal CLOCK_NAME: MVL)
                         return BOOLEAN;
  function ES_FALLING    (signal CLOCK_NAME: MVL)
                         return BOOLEAN;

  -- Resolution functions:
  function WIRED_AND (SIG_DRIVERS: MVL_VECTOR)
                         return MVL;

  function WIRED_OR      (SIG_DRIVERS: MVL_VECTOR)
                         return MVL;
end ATT_MVL;
```

```
package body ATT_MVL is
  -- Function definitions for overloaded operators with MVL types:
  function "and" (L, R: MVL) return MVL is
  begin
    return TABLE_AND(L, R);
  end "and";
  function "or" (L, R: MVL) return MVL is
  begin
    return TABLE_OR(L, R);
  end "or";
  function "nand" (L, R: MVL) return MVL is
  begin
    return TABLE_NAND(L, R);
  end "nand";
  function "nor" (L, R: MVL) return MVL is
  begin
    return TABLE_NOR(L, R);
  end "nor";
  function "xor" (L, R: MVL) return MVL is
  begin
    return TABLE_XOR(L, R);
  end "xor";
  function "not" (L: MVL) return MVL is
  begin
    return TABLE_NOT(L);
  end "not";

  -- Function definitions for MVL_VECTOR types:
  function "and" (L, R: MVL_VECTOR) return MVL_VECTOR is
    alias LOP: MVL_VECTOR (L'LENGTH–1 downto 0) is L;
    alias ROP: MVL_VECTOR (R'LENGTH–1 downto 0) is R;
    variable RESULT: MVL_VECTOR (L'LENGTH–1 downto 0);
  begin
    if L'LENGTH /= R'LENGTH then
      report "Operands of overloaded AND operator " &
             " are not of same length!"
        severity FAILURE;
      return RESULT;
    end if;
    for K in RESULT'RANGE loop
      RESULT(K) := TABLE_AND (LOP(K), ROP(K));
    end loop;
    return RESULT;
  end "and";
  function "or" (L, R: MVL_VECTOR) return MVL_VECTOR is
    alias LOP: MVL_VECTOR (L'LENGTH–1 downto 0) is L;
    alias ROP: MVL_VECTOR (R'LENGTH–1 downto 0) is R;
```

```
    variable RESULT: MVL_VECTOR (L'LENGTH–1 downto 0);
begin
    if L'LENGTH /= R'LENGTH then
        report "Operands of overloaded OR operator " &
                " are not of same length!"
            severity FAILURE;
        return RESULT;
    end if;
    for K in RESULT'RANGE loop
        RESULT(K) := TABLE_OR (LOP(K), ROP(K));
    end loop;
    return RESULT;
end "or";
function "nand" (L, R: MVL_VECTOR) return MVL_VECTOR is
    alias LOP: MVL_VECTOR (L'LENGTH–1 downto 0) is L;
    alias ROP: MVL_VECTOR (R'LENGTH–1 downto 0) is R;
    variable RESULT: MVL_VECTOR (L'LENGTH–1 downto 0);
begin
    if L'LENGTH /= R'LENGTH then
        report "Operands of overloaded NAND operator " &
                " are not of same length!"
            severity FAILURE;
        return RESULT;
    end if;
    for K in RESULT'RANGE loop
        RESULT(K) := TABLE_NAND (LOP(K), ROP(K));
    end loop;
    return RESULT;
end "nand";
function "nor" (L, R: MVL_VECTOR) return MVL_VECTOR is
    alias LOP: MVL_VECTOR (L'LENGTH–1 downto 0) is L;
    alias ROP: MVL_VECTOR (R'LENGTH–1 downto 0) is R;
    variable RESULT: MVL_VECTOR (L'LENGTH–1 downto 0);
begin
    if L'LENGTH /= R'LENGTH then
        report "Operands of overloaded NOR operator " &
                " are not of same length!"
            severity FAILURE;
        return RESULT;
    end if;
    for K in RESULT'RANGE loop
        RESULT(K) := TABLE_NOR (LOP(K), ROP(K));
    end loop;
    return RESULT;
end "nor";
function "xor" (L, R: MVL_VECTOR) return MVL_VECTOR is
    alias LOP: MVL_VECTOR (L'LENGTH–1 downto 0) is L;
```

```
    alias ROP: MVL_VECTOR (R'LENGTH–1 downto 0) is R;
    variable RESULT: MVL_VECTOR (L'LENGTH–1 downto 0);
  begin
    if L'LENGTH /= R'LENGTH then
      report "Operands of overloaded XOR operator " &
             " are not of same length!"
        severity FAILURE;
      return RESULT;
    end if;
    for K in RESULT'RANGE loop
      RESULT(K) := TABLE_XOR (LOP(K), ROP(K));
    end loop;
    return RESULT;
  end "xor";
  function "not" (L: MVL_VECTOR) return MVL_VECTOR is
    alias LOP: MVL_VECTOR (L'LENGTH–1 downto 0) is L;
    variable RESULT: MVL_VECTOR (L'LENGTH–1 downto 0);
  begin
    for K in RESULT'RANGE loop
      RESULT(K) := TABLE_NOT (LOP(K));
    end loop;
    return RESULT;
  end "not";

  -- Common utilities:
  function MAX  (T1, T2: TIME) return TIME is
  begin
    if T1 > T2 then
      return T1;
    else
      return T2;
    end if;
  end MAX;
  function INT2MVL (OPD, NO_BITS: INTEGER)
                            return MVL_VECTOR is
    variable M1: INTEGER;
    variable RET: MVL_VECTOR (NO_BITS–1 downto 0)
           := (others => '0');
  begin
    if OPD < 0 then
      report "Operand is less than 0!"
        severity FAILURE;
      return RET;
    end if;
    M1 := OPD;
    for J in RET'REVERSE_RANGE loop
      if (M1 mod 2) = 1 then
```

```
            RET(J) := '1';
        else
            RET(J) := '0';
        end if;
        M1 := M1 / 2;
    end loop;
    return RET;
end INT2MVL;
function MVL2INT (OPD: MVL_VECTOR) return INTEGER is
    -- Leftmost is MSB.
    variable TEMP: NATURAL := 0;
    variable J: NATURAL := OPD'LENGTH – 1;
begin
    if OPD'LENGTH >= 32 then
        report "Operand has 32 bits or more!"
            severity FAILURE;
        return TEMP;
    end if;
    for M3 in OPD'RANGE loop
        if OPD(M3) = '1' then
            TEMP := TEMP + 2 ** J;
        end if;
        J := J – 1;
    end loop;
    return TEMP;
end MVL2INT;
procedure P_1 (signal DRIVE_SIGNAL: inout MVL;
                    SIGNAL_VALUE: in MVL;
                    DRIVE_WIDTH, DRIVE_DELAY: in TIME) is
begin
    if SIGNAL_VALUE = '0' then
        DRIVE_SIGNAL <= '0';
    else
        DRIVE_SIGNAL <= '1' after DRIVE_DELAY,
            '0' after DRIVE_WIDTH+DRIVE_DELAY;
    end if;
end P_1;

-- Overloaded arithmetic operators on MVL_VECTOR:
function "+"    (L, R: MVL_VECTOR) return MVL_VECTOR is
    alias LOP: MVL_VECTOR (L'LENGTH–1 downto 0) is L;
    alias ROP: MVL_VECTOR (R'LENGTH–1 downto 0) is R;
    -- Assume 0 is LSB; unsigned numbers.
    variable SUM: MVL_VECTOR (L'LENGTH–1 downto 0);
    variable CARRY: MVL := '0';
begin
    if L'LENGTH /= R'LENGTH then
```

```
      report "Operands of overloaded + operator " &
              " are not of same length!"
        severity FAILURE;
      return SUM;
    end if;
    for J in SUM'REVERSE_RANGE loop
      SUM(J) := LOP(J) xor ROP(J) xor CARRY;
      CARRY := (LOP(J) and ROP(J)) or (LOP(J) and CARRY)
                 or (ROP(J) and CARRY);
    end loop;
    return SUM;
  end "+";
  function "*"      (L, R: MVL_VECTOR) return MVL_VECTOR is
    -- Unsigned numbers being multiplied; result < 32 bits.
    variable T1, T2: NATURAL;
    constant R_SIZE: NATURAL := L'LENGTH + R'LENGTH;
    variable RESULT: MVL_VECTOR (R_SIZE–1 downto 0);
  begin
    if L'LENGTH+R'LENGTH >= 32 then
      report "Total number of bits in operands is 32 or more!"
        severity FAILURE;
      return RESULT;
    end if;
    T1 := MVL2INT (L);
    T2 := MVL2INT (R);
    RESULT := INT2MVL (T1 * T2, R_SIZE);
    return RESULT;
  end "*";

  -- Clock functions:
  function ES_RISING (signal CLOCK_NAME: MVL)
                        return BOOLEAN is
  begin
    return CLOCK_NAME = '1' and CLOCK_NAME'EVENT and
        CLOCK_NAME'LAST_VALUE = '0';
  end ES_RISING;
  function ES_FALLING (signal CLOCK_NAME: MVL)
                        return BOOLEAN is
  begin
    return CLOCK_NAME = '0' and CLOCK_NAME'EVENT and
         CLOCK_NAME'LAST_VALUE = '1';
  end ES_FALLING;

  -- Resolution functions:
  function WIRED_AND      (SIG_DRIVERS: MVL_VECTOR)
                        return MVL is
    constant MEMORY: MVL := 'Z';
```

```
    variable RESOLVE_VALUE: MVL;
    variable FIRST: BOOLEAN := TRUE;
  begin
    if SIG_DRIVERS'LENGTH = 0 then
      return MEMORY;
    else
      for K in SIG_DRIVERS'RANGE loop
        if FIRST then
          RESOLVE_VALUE := SIG_DRIVERS(K);
          FIRST := FALSE;
        else
          RESOLVE_VALUE := TABLE_TAND
              (RESOLVE_VALUE, SIG_DRIVERS(K));
        end if;
      end loop;
      return RESOLVE_VALUE;
    end if;
  end WIRED_AND;
  function WIRED_OR      (SIG_DRIVERS: MVL_VECTOR)
                         return MVL is
    constant MEMORY: MVL := 'Z';
    variable RESOLVE_VALUE: MVL;
    variable FIRST: BOOLEAN := TRUE;
  begin
    if SIG_DRIVERS'LENGTH = 0 then
      return MEMORY;
    else
      for K in SIG_DRIVERS'RANGE loop
        if FIRST then
          RESOLVE_VALUE := SIG_DRIVERS(K);
          FIRST := FALSE;
        else
          RESOLVE_VALUE := TABLE_TOR
              (RESOLVE_VALUE, SIG_DRIVERS(K));
        end if;
      end loop;
      return RESOLVE_VALUE;
    end if;
  end WIRED_OR;
end ATT_MVL;
```

❑

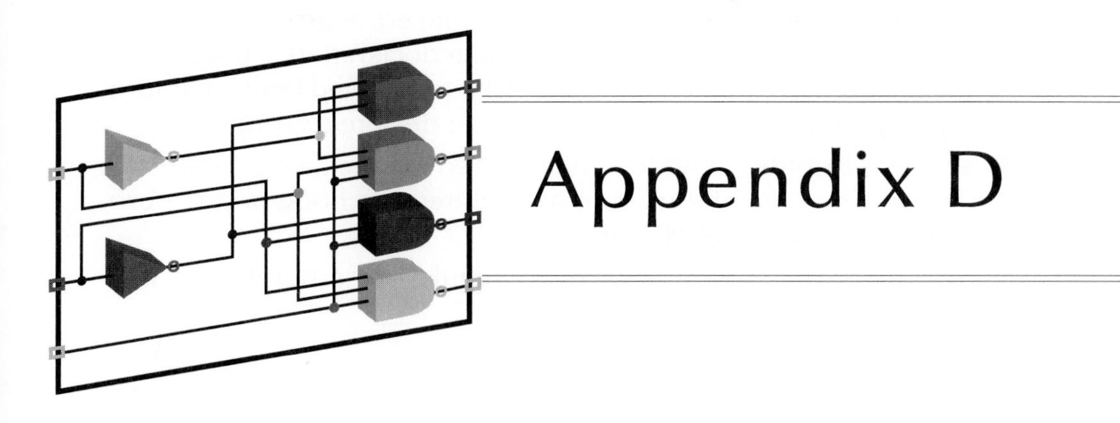

Appendix D

Summary of Changes

This appendix presents the changes that were introduced to the 1076-1987 version of the language. It also contains a section on the portability of models from the 1076-1987 to 1076-1993 versions of the language.

D.1 VHDL'93 Features

This section lists the key features that have been added to the 1076-1993 language. In addition to those listed below, semantics of a number of constructs have also been clarified.

1. File is the fourth object class, in addition to variable, constant, and signal.
2. A constant expression is allowed as an actual in a port map.
3. Shared variables are allowed, that is, variables can be declared outside of a process or a subprogram. Thus, more than one process may read or write to a shared variable.

4. The notion of a group has been introduced. A group is a collection of named items. A group template declaration is used to declare a group template, that is, the class of named items that form a group. A group declaration is used to declare a group.
5. A new attribute 'FOREIGN is declared in package STANDARD. This attribute can be used in an architecture body or a subprogram to link in non-VHDL models.
6. The syntax has been made more regular for a number of constructs.

```
component C is
   ...
end component C;

process (...) is
   ...
end process;

A: block (...) is
   ...
end block A;

architecture A of E is
   ...
end architecture A;

procedure P is
   ...
end procedure P;

function F (...) is
   ...
end function F;

entity E is
   ...
end entity E;

configuration C of E is
   ...
end configuration C;

package PK is
   ...
end package PK;

package body PK is
   ...
end package body PK;
```

7. All sequential statements can be labeled. For example,

```
L1: if A = B then            -- L1 is the if statement label.
   C := D;
end if L1;                   -- Label can optionally appear at the end.

L2: SUM := (A xor B) xor C;
```

8. Functions can be designated as pure and impure. A pure function is one that returns the same value for multiple calls to the function with the same set of parameter values; an impure function may return a different value when called multiple times, even with the same parameter values.
9. The notion of signature has been introduced. This can be used to explicitly identify overloaded subprograms and overloaded enumeration literals. The signature explicitly specifies the parameter and result profile.
10. File operations that are implicitly defined when a file type is declared have been redefined. These are FILE_OPEN, FILE_CLOSE, READ, WRITE, and ENDFILE.
11. The syntax of file declaration has been redefined.
12. An alias can be specified for any named item, that is, an alias is not restricted to just objects. However, labels, and loop and generate parameters, cannot be aliased.
13. An attribute can be specified for a literal, units, a group, and a file.
14. The `xnor` operator and shift and rotate operators have been defined.
15. The results of a concatenation operator have been redefined.
16. The report statement has been introduced. It is very similar to an assertion statement but without the assert expression.
17. A pulse rejection window can be specified when inertial delay is used in a signal assignment.
18. A value of **unaffected** can be assigned to a signal to indicate no change to the value of the driver.
19. A process can be marked as a postponed process. A postponed process executes only at the end of a time step, that is, after all the deltas of a time step.
20. Similarly, a concurrent assertion statement, a concurrent procedure call, and a concurrent signal assignment statement can be marked as postponed as well.

21. Direct instantiation is allowed; that is, in a component instantiation statement, an entity-architecture pair or a configuration can be directly instantiated.
22. Incremental binding is allowed. For example, a configuration specification in an architecture body may specify the binding to a design entity without specifying the port map and the generic map, while a configuration declaration may be used later to specify the port map and the generic map.
23. The generate statement can have a declarative part.
24. The character set has been extended to include a number of other special characters.
25. Extended identifers have been defined. An extended identifier is a sequence of characters written between two backslashes.
26. A bit string literal represents a sequence of bits. The type of literal need not necessarily be BIT_VECTOR; the type is determined from the context in which the literal appears.
27. The following predefined attributes have been added:

    'ASCENDING, 'IMAGE, 'VALUE, 'DRIVING, 'DRIVING_VALUE, 'SIMPLE_NAME, 'INSTANCE_NAME, 'PATH_NAME

    The following predefined attributes have been deleted:

    'STRUCTURE, 'BEHAVIOR.
28. In package STANDARD, the following have been added:

    - DELAY_LENGTH physical subtype
    - FILE_OPEN_KIND enumeration type
    - FILE_OPEN_STATUS enumeration type
    - FOREIGN attribute declaration
29. The semantics of subprograms in the TEXTIO package have been elaborated.

D.2 Portability from VHDL'87

This section describes some of the features changed in 1076-1993 that can cause models written in 1076-1987 to be non-portable.

1. Enumeration type CHARACTER has a much larger set. Therefore, code that depended on 'HIGH, 'RIGHT of type CHARACTER might find some changes.
2. Ambiguities in the usage of the concatenation operator have been clarified.
3. File type declaration and file declarations have been redefined, and their associated implicit file operations have also been redefined.
4. 'STRUCTURE and 'BEHAVIOR attributes are no longer present in the language. Therefore, any code that uses these attributes will need to be changed.
5. New reserved words have been added to the language. Therefore, if these were used as identifiers, the identifiers will have to be changed.

❑

D.2 Portability from VHDL'87

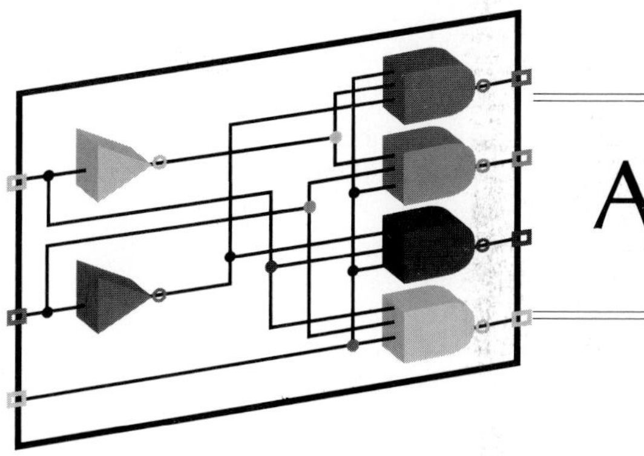

Appendix E

The STD_LOGIC_1164 Package

This appendix provides a listing of the IEEE standard multivalue logic system for VHDL model interoperability, the STD_LOGIC_1164 package. This package is used in many examples in this book. [1]

E.1 Package STD_LOGIC_1164

```
-- This package shall be compiled into a design library
-- symbolically named IEEE.

package STD_LOGIC_1164 is
    -------------------------------------------------------------
    -- Logic State System (unresolved)
    -------------------------------------------------------------
```

1. Reprinted from IEEE Std 1164-1993 with permission from IEEE.

```
type STD_ULOGIC is ( 'U',   -- Uninitialized
                     'X',   -- Forcing Unknown
                     '0',   -- Forcing 0
                     '1',   -- Forcing 1
                     'Z',   -- High Impedance
                     'W',   -- Weak   Unknown
                     'L',   -- Weak   0
                     'H',   -- Weak   1
                     '-'    -- don't care
                   );
-------------------------------------------------------------------
-- Unconstrained array of std_ulogic for use with the
-- resolution function
-------------------------------------------------------------------
type STD_ULOGIC_VECTOR is array ( NATURAL range <> )
    of STD_ULOGIC;
-------------------------------------------------------------------
-- resolution function
-------------------------------------------------------------------
function RESOLVED ( S : STD_ULOGIC_VECTOR )
                        return STD_ULOGIC;
-------------------------------------------------------------------
-- *** industry standard logic type ***
-------------------------------------------------------------------
subtype STD_LOGIC is RESOLVED STD_ULOGIC;
-------------------------------------------------------------------
-- Unconstrained array of std_logic for use in declaring
-- signal arrays
-------------------------------------------------------------------
type STD_LOGIC_VECTOR is array ( NATURAL range <> )
    of STD_LOGIC;
-------------------------------------------------------------------
-- common subtypes
-------------------------------------------------------------------
subtype X01 is RESOLVED STD_ULOGIC range 'X' to '1';
  -- ('X', '0', '1')
subtype X01Z is RESOLVED STD_ULOGIC range 'X' to 'Z';
  -- ('X', '0', '1', 'Z')
subtype UX01 is RESOLVED STD_ULOGIC range 'U' to '1';
  -- ('U', 'X', '0', '1')
subtype UX01Z  is RESOLVED STD_ULOGIC range 'U' to 'Z';
  -- ('U', 'X', '0', '1', 'Z')
-------------------------------------------------------------------
-- overloaded logical operators
-------------------------------------------------------------------
function "and" ( L : STD_ULOGIC; R : STD_ULOGIC ) return UX01;
function "nand" ( L : STD_ULOGIC; R : STD_ULOGIC )
```

```
                  return UX01;
function "or" ( L : STD_ULOGIC; R : STD_ULOGIC ) return UX01;
function "nor" ( L : STD_ULOGIC; R : STD_ULOGIC ) return UX01;
function "xor" ( L : STD_ULOGIC; R : STD_ULOGIC ) return UX01;
function "xnor" ( L : STD_ULOGIC; R : STD_ULOGIC )
                  return UX01;
function "not" ( L : STD_ULOGIC ) return UX01;
-------------------------------------------------------------------
-- vectorized overloaded logical operators
-------------------------------------------------------------------
function "and"  ( L, R : STD_LOGIC_VECTOR )
                         return STD_LOGIC_VECTOR;
function "and"  ( L, R : STD_ULOGIC_VECTOR )
                         return STD_ULOGIC_VECTOR;
function "nand" ( L, R : STD_LOGIC_VECTOR )
                           return STD_LOGIC_VECTOR;
function "nand" ( L, R : STD_ULOGIC_VECTOR )
                           return STD_ULOGIC_VECTOR;
function "or"   ( L, R : STD_LOGIC_VECTOR )
                         return STD_LOGIC_VECTOR;
function "or"   ( L, R : STD_ULOGIC_VECTOR )
                         return STD_ULOGIC_VECTOR;
function "nor"  ( L, R : STD_LOGIC_VECTOR )
                         return STD_LOGIC_VECTOR;
function "nor"  ( L, R : STD_ULOGIC_VECTOR )
                         return STD_ULOGIC_VECTOR;
function "xor"  ( L, R : STD_LOGIC_VECTOR )
                         return STD_LOGIC_VECTOR;
function "xor"  ( L, R : STD_ULOGIC_VECTOR )
                         return STD_ULOGIC_VECTOR;
function "xnor" ( L, R : STD_LOGIC_VECTOR )
                         return STD_LOGIC_VECTOR;
function "xnor" ( L, R : STD_ULOGIC_VECTOR )
                         return STD_ULOGIC_VECTOR;
function "not"  ( L    : STD_LOGIC_VECTOR )
                         return STD_LOGIC_VECTOR;
function "not"  ( L    : STD_ULOGIC_VECTOR )
                         return STD_ULOGIC_VECTOR;
-------------------------------------------------------------------
-- conversion functions
-------------------------------------------------------------------
function TO_BIT ( S : STD_ULOGIC; XMAP : BIT := '0')
                   return BIT;
function TO_BITVECTOR     ( S : STD_LOGIC_VECTOR;
                           XMAP : BIT := '0') return BIT_VECTOR;
function TO_BITVECTOR     ( S : STD_ULOGIC_VECTOR;
                           XMAP : BIT := '0') return BIT_VECTOR;
```

```
function TO_STDULOGIC        (B : BIT ) return STD_ULOGIC;
function TO_STDLOGICVECTOR     ( B : BIT_VECTOR      )
                               return STD_LOGIC_VECTOR;
function TO_STDLOGICVECTOR     ( S : STD_ULOGIC_VECTOR )
                               return STD_LOGIC_VECTOR;
function TO_STDULOGICVECTOR    ( B : BIT_VECTOR      )
                               return STD_ULOGIC_VECTOR;
function TO_STDULOGICVECTOR    ( S : STD_LOGIC_VECTOR )
                               return STD_ULOGIC_VECTOR;
-------------------------------------------------------------------
-- strength strippers and type convertors
-------------------------------------------------------------------
function TO_X01  ( S : STD_LOGIC_VECTOR )
                       return STD_LOGIC_VECTOR;
function TO_X01  ( S : STD_ULOGIC_VECTOR)
                       return STD_ULOGIC_VECTOR;
function TO_X01  ( S : STD_ULOGIC      ) return X01;
function TO_X01  ( B : BIT_VECTOR      )
                       return STD_LOGIC_VECTOR;
function TO_X01  ( B : BIT_VECTOR      )
                       return STD_ULOGIC_VECTOR;
function TO_X01  ( B : BIT             ) return X01;
function TO_X01Z ( S : STD_LOGIC_VECTOR )
                       return STD_LOGIC_VECTOR;
function TO_X01Z ( S : STD_ULOGIC_VECTOR)
                       return STD_ULOGIC_VECTOR;
function TO_X01Z ( S : STD_ULOGIC      ) return X01Z;
function TO_X01Z ( B : BIT_VECTOR      )
                       return STD_LOGIC_VECTOR;
function TO_X01Z ( B : BIT_VECTOR      )
                       return STD_ULOGIC_VECTOR;
function TO_X01Z ( B : BIT             ) return X01Z;
function TO_UX01 ( S : STD_LOGIC_VECTOR )
                       return STD_LOGIC_VECTOR;
function TO_UX01 ( S : STD_ULOGIC_VECTOR)
                       return STD_ULOGIC_VECTOR;
function TO_UX01 ( S : STD_ULOGIC      ) return UX01;
function TO_UX01 ( B : BIT_VECTOR      )
                       return STD_LOGIC_VECTOR;
function TO_UX01 ( B : BIT_VECTOR      )
                       return STD_ULOGIC_VECTOR;
function TO_UX01 ( B : BIT             ) return UX01;
-------------------------------------------------------------------
-- edge detection
-------------------------------------------------------------------
function RISING_EDGE  (signal S : STD_ULOGIC)
                              return BOOLEAN;
```

```
function FALLING_EDGE (signal S : STD_ULOGIC)
                                 return BOOLEAN;
-------------------------------------------------------------------
-- object contains an unknown
-------------------------------------------------------------------
function IS_X ( S : STD_ULOGIC_VECTOR ) return  BOOLEAN;
function IS_X ( S : STD_LOGIC_VECTOR  ) return  BOOLEAN;
function IS_X ( S : STD_ULOGIC        ) return  BOOLEAN;

end STD_LOGIC_1164;
```

❑

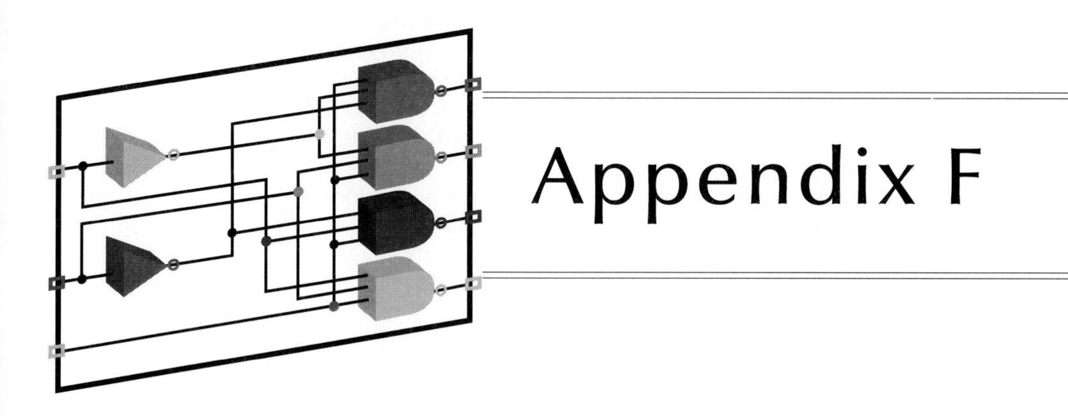

Appendix F

An Utility Package

This appendix describes the package UTILS_PKG that is used in some examples of this book.

F.1 Package UTILS_PKG

```
-- In examples in this book wherever used, it is assumed that
-- this package is compiled into a design library called BOOK_LIB.

library IEEE;
use IEEE.STD_LOGIC_1164.all;
package UTILS_PKG is
  -- A STD_LOGIC_VECTOR is interpreted as an unsigned value and
  -- it's integer value is returned:
  function TO_INTEGER (OPD: STD_LOGIC_VECTOR)
                        return INTEGER;
  -- A non-negative integer is converted to a vector form of
  -- the specified size:
  function TO_STDLOGICVECTOR
                        (OPD: NATURAL; NO_BITS: NATURAL)
                        return STD_LOGIC_VECTOR;
  function TO_STDLOGICVECTOR (V: STRING)
```

```
                                   return STD_LOGIC_VECTOR;
  function TO_STRING (V: STD_LOGIC_VECTOR) return STRING;
  function TO_STDLOGIC (V: CHARACTER) return STD_LOGIC;
  function TO_CHARACTER (V: STD_LOGIC)
                              return CHARACTER;

  -- Function returns the maximum of the two time values:
  function MAX      (T1, T2: TIME) return TIME;

  -- Generate a pulse of specified value and width after
  -- specified delay:
  procedure GENERATE_PULSE
                        (signal DRIVE_SIGNAL: inout STD_LOGIC;
                        SIGNAL_VALUE: in STD_LOGIC;
                        DRIVE_WIDTH, DRIVE_DELAY: in TIME);

  -- Overloaded arithmetic operators on STD_LOGIC_VECTOR.
  -- The vectors are interpreted as unsigned numbers:
  function "+"    (L, R: STD_LOGIC_VECTOR)
                  return STD_LOGIC_VECTOR;
  function "*"    (L, R: STD_LOGIC_VECTOR)
                  return STD_LOGIC_VECTOR;

  -- Resolution functions:
  function WIRED_AND (SIG_DRIVERS: STD_LOGIC_VECTOR)
                       return STD_LOGIC;
  function WIRED_OR    (SIG_DRIVERS: STD_LOGIC_VECTOR)
                       return STD_LOGIC;
  function WIRED_AND (SIG_DRIVERS: BIT_VECTOR)
                       return BIT;
  function WIRED_OR    (SIG_DRIVERS: BIT_VECTOR)
                       return BIT;
end UTILS_PKG;

package body UTILS_PKG is
  function TO_INTEGER (OPD: STD_LOGIC_VECTOR)
                       return INTEGER is
    -- Leftmost is MSB.
    variable TEMP: NATURAL := 0;
    variable J: NATURAL := OPD'LENGTH – 1;
  begin
    if OPD'LENGTH >= 32 then
      report "Operand has 32 bits or more!"
        severity FAILURE;
      return TEMP;
    end if;
    for M3 in OPD'RANGE loop
      if OPD(M3) = '1' then
```

```
      TEMP := TEMP + 2 ** J;
    end if;
    J := J – 1;
  end loop;
  return TEMP;
end TO_INTEGER;

function TO_STDLOGICVECTOR
                        (OPD: NATURAL; NO_BITS: NATURAL)
                        return STD_LOGIC_VECTOR is
  variable M1: INTEGER;
  variable RET: STD_LOGIC_VECTOR (NO_BITS–1 downto 0)
          := (others => '0');
begin
  M1 := OPD;
  for J in RET'REVERSE_RANGE loop
    if (M1 mod 2) = 1 then
      RET(J) := '1';
    else
      RET(J) := '0';
    end if;
    M1 := M1 / 2;
  end loop;
  return RET;
end TO_STDLOGICVECTOR;

function TO_STDLOGIC (V: CHARACTER) return STD_LOGIC is
begin
  case V is
    when 'X' => return 'X';
    when '0' => return '0';
    when '1' => return '1';
    when 'Z' => return 'Z';
    when 'U' => return 'U';
    when 'W' => return 'W';
    when 'L' => return 'L';
    when 'H' => return 'H';
    when '-' => return '-';
    when others =>
      assert FALSE
        report "A character other than U,X,Z,0,1,W,L,H,- found"
        severity ERROR;
      return '-';
  end case;
end TO_STDLOGIC;

function TO_CHARACTER (V: STD_LOGIC)
```

```
                                        return CHARACTER is
begin
  case V is
    when 'X' => return 'X';
    when '0' => return '0';
    when '1' => return '1';
    when 'Z' => return 'Z';
    when 'U' => return 'U';
    when 'W' => return 'W';
    when 'L' => return 'L';
    when 'H' => return 'H';
    when '-' => return '-';
  end case;
end TO_CHARACTER;

function TO_STRING (V: STD_LOGIC_VECTOR)
                              return STRING is
  variable RET: STRING (1 to V'LENGTH);
  variable K: INTEGER := 1;
begin
  for J in V'RANGE loop
    RET(K) := TO_CHARACTER(V(J));
    K := K + 1;
  end loop;
  return RET;
end TO_STRING;

function TO_STDLOGICVECTOR (V: STRING)
                                        return STD_LOGIC_VECTOR is
  variable RET : STD_LOGIC_VECTOR(0 to V'LENGTH-1);
  variable K: INTEGER;          -- The range for STRING starts from 1.
begin
  for J in V'RANGE loop
    K := J - 1;
    RET(K) := TO_STDLOGIC(V(J));
  end loop;
  return RET;
end TO_STDLOGICVECTOR;

function MAX      (T1, T2: TIME) return TIME is
begin
  if T1 > T2 then
    return T1;
  else
    return T2;
  end if;
end MAX;
```

```
procedure GENERATE_PULSE
                    (signal DRIVE_SIGNAL: inout STD_LOGIC;
                    SIGNAL_VALUE: in STD_LOGIC;
                    DRIVE_WIDTH, DRIVE_DELAY: in TIME) is
begin
  if SIGNAL_VALUE = '0' then
    DRIVE_SIGNAL <= '0';
  else
    DRIVE_SIGNAL <= '1' after DRIVE_DELAY,
      '0' after DRIVE_WIDTH+DRIVE_DELAY;
  end if;
end GENERATE_PULSE;

function "+"    (L, R: STD_LOGIC_VECTOR)
                return STD_LOGIC_VECTOR is
  alias LOP: STD_LOGIC_VECTOR (L'LENGTH–1 downto 0) is L;
  alias ROP: STD_LOGIC_VECTOR (R'LENGTH–1 downto 0) is R;
  -- Assume 0 is LSB; unsigned numbers.
  variable SUM: STD_LOGIC_VECTOR (L'LENGTH–1 downto 0);
  variable CARRY: STD_LOGIC := '0';
begin
  if L'LENGTH /= R'LENGTH then
    report "Operands of overloaded + operator " &
           " are not of same length!"
      severity FAILURE;
    return SUM;
  end if;
  for J in SUM'REVERSE_RANGE loop
    SUM(J) := LOP(J) xor ROP(J) xor CARRY;
    CARRY := (LOP(J) and ROP(J)) or (LOP(J) and CARRY)
                or (ROP(J) and CARRY);
  end loop;
  return SUM;
end "+";

function "*"    (L, R: STD_LOGIC_VECTOR)
                return STD_LOGIC_VECTOR is
  -- Unsigned numbers being multiplied; result < 32 bits.
  variable T1, T2: NATURAL;
  constant R_SIZE: NATURAL := L'LENGTH + R'LENGTH;
  variable RESULT: STD_LOGIC_VECTOR (R_SIZE–1 downto 0);
begin
  if L'LENGTH+R'LENGTH >= 32 then
    report "Total number of bits in operands is 32 or more!"
      severity FAILURE;
    return RESULT;
```

```
    end if;
    T1 := TO_INTEGER (L);
    T2 := TO_INTEGER (R);
    RESULT := TO_STDLOGICVECTOR (T1 * T2, R_SIZE);
    return RESULT;
  end "*";

  function WIRED_AND (SIG_DRIVERS: STD_LOGIC_VECTOR)
                          return STD_LOGIC is
    constant MEMORY: STD_LOGIC := 'Z';
    variable RESOLVE_VALUE: STD_LOGIC;
    variable FIRST: BOOLEAN := TRUE;
  begin
    if SIG_DRIVERS'LENGTH = 0 then
      return MEMORY;
    else
      for K in SIG_DRIVERS'RANGE loop
        if FIRST then
          RESOLVE_VALUE := SIG_DRIVERS(K);
          FIRST := FALSE;
        else
          RESOLVE_VALUE :=
                        RESOLVE_VALUE and SIG_DRIVERS(K);
        end if;
      end loop;
      return RESOLVE_VALUE;
    end if;
  end WIRED_AND;

  function WIRED_OR      (SIG_DRIVERS: STD_LOGIC_VECTOR)
      return STD_LOGIC is
    constant MEMORY: STD_LOGIC := 'Z';
    variable RESOLVE_VALUE: STD_LOGIC;
    variable FIRST: BOOLEAN := TRUE;
  begin
    if SIG_DRIVERS'LENGTH = 0 then
      return MEMORY;
    else
      for K in SIG_DRIVERS'RANGE loop
        if FIRST then
          RESOLVE_VALUE := SIG_DRIVERS(K);
          FIRST := FALSE;
        else
          RESOLVE_VALUE :=
                          RESOLVE_VALUE or SIG_DRIVERS(K);
        end if;
      end loop;
```

```
    return RESOLVE_VALUE;
  end if;
end WIRED_OR;

function WIRED_AND (SIG_DRIVERS: BIT_VECTOR)
    return BIT is
  constant MEMORY: BIT := '0';
  variable RESOLVE_VALUE: BIT;
  variable FIRST: BOOLEAN := TRUE;
begin
  if SIG_DRIVERS'LENGTH = 0 then
    return MEMORY;
  else
    for K in SIG_DRIVERS'RANGE loop
      if FIRST then
        RESOLVE_VALUE := SIG_DRIVERS(K);
        FIRST := FALSE;
      else
        RESOLVE_VALUE :=
                         RESOLVE_VALUE and SIG_DRIVERS(K);
      end if;
    end loop;
    return RESOLVE_VALUE;
  end if;
end WIRED_AND;

function WIRED_OR       (SIG_DRIVERS: BIT_VECTOR)
    return BIT is
  constant MEMORY: BIT := '0';
  variable RESOLVE_VALUE: BIT;
  variable FIRST: BOOLEAN := TRUE;
begin
  if SIG_DRIVERS'LENGTH = 0 then
    return MEMORY;
  else
    for K in SIG_DRIVERS'RANGE loop
      if FIRST then
        RESOLVE_VALUE := SIG_DRIVERS(K);
        FIRST := FALSE;
      else
        RESOLVE_VALUE :=
                         RESOLVE_VALUE or SIG_DRIVERS(K);
      end if;
    end loop;
    return RESOLVE_VALUE;
  end if;
end WIRED_OR;
```

```
end UTILS_PKG;
```

❑

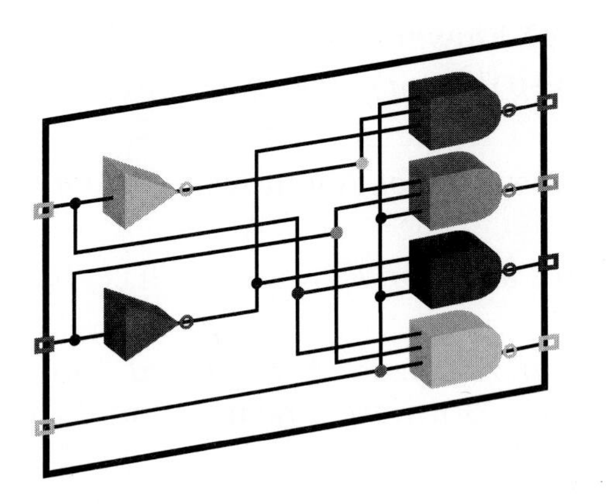

Bibliography

Following is a list of suggested readings and books on the language. The list is not intended to be comprehensive.

1. Armstrong, J. R., *Chip-level Modeling with VHDL*, Englewood Cliffs, NJ: Prentice Hall, 1988.
2. Armstrong, J. R. et al., *The VHDL validation suite*, Proc. 27th Design Automation Conference, June 1990, pp. 2-7.
3. Ashenden, P. J., *The Designers Guide to VHDL*, Morgan Kaufmann, 1994.
4. Ashenden, P. J., *The VHDL Cookbook*, University of Adelaide, Australia, 1990.
5. Baker, L., *VHDL Programming with Advanced Topics*, John Wiley and Sons, Inc., 1993.
6. Barton, D., *A first course in VHDL*, VLSI Systems Design, January 1988.
7. Berge, J.-M. et al., *VHDL Designer's Reference*, Kluwer Academic, 1992.

8. Berge, J.-M. et al., *VHDL '92*, Kluwer Academic, 1993.

9. Bhasker, J., *A Guide to VHDL Syntax*, Englewood Cliffs, NJ: Prentice Hall, 1994.

10. Bhasker, J., *A VHDL Synthesis Primer*, Allentown, PA: Star Galaxy Publishing, 1995.

11. Bhasker, J., *VHDL: Features and Applications*, NJ: IEEE, Order No. HL5712, 1996.

12. Bhasker, J., *Process-Graph Analyzer: A front-end tool for VHDL behavioral synthesis*, Software Practice and Experience, vol. 18, no. 5, May 1988.

13. Bhasker, J., *An algorithm for microcode compaction of VHDL behavioral descriptions*, Proc. 20th Microprogramming Workshop, December 1987.

14. Coelho, D., *The VHDL Handbook*, Boston: Kluwer Academic, 1988.

15. Coelho, D., *VHDL: A call for standards*, Proc. 25th Design Automation Conference, June 1988.

16. Farrow, R., and A. Stanculescu, *A VHDL compiler based on attribute grammar methodology*, SIGPLAN 1989.

17. Gilman, A. S., *Logic Modeling in WAVES*, IEEE Design & Test of Computers, June 1990, pp. 49-55.

18. Hands, J. P., *What is VHDL?* Computer-Aided Design, vol. 22, no. 4, May 1990.

19. Harr, R., and A. Stanculescu (eds.), *Applications of VHDL to Circuit Design*, Boston: Kluwer Academic, 1991.

20. Hines, J., *Where VHDL fits within the CAD environment*, Proc. 24th Design Automation Conference, 1987.

21. *IEEE Standard VHDL Language Reference Manual, Std 1076-1987*, IEEE, NY, 1988.

22. *IEEE Standard VHDL Language Reference Manual, Std 1076-1993*, IEEE, NY, 1993.

23. *IEEE Standard 1076 VHDL Tutorial*, CLSI, Maryland, March 1989.

24. *IEEE Standard Interpretations: IEEE Std 1076-1987, IEEE Standard VHDL Language Reference Manual*, IEEE, 1992.

25. *IEEE Standard Multivalue Logic System for VHDL Model Interoperability (Std_Logic_1164)*, Std 1164-1993, IEEE, 1993.
26. Kim, K., and J. Trout, *Automatic insertion of BIST hardware using VHDL*, Proc. 25th Design Automation Conference, 1988.
27. Leung, *ASIC System Design with VHDL*, Boston: Kluwer Academic, 1989.
28. Lipsett, R. et. al., *VHDL: Hardware Description and Design*, Boston: Kluwer Academic, 1989.
29. Moughzail, M. et. al., *Experience with the VHDL environment*, Proc. 25th Design Automation Conference, 1988.
30. Perry, D., *VHDL*, New York: McGraw Hill, 1991.
31. *Military Standard 454*, U.S. Government Printing Office, 1988.
32. Navabi, Z., *VHDL Analysis and Modeling of Digital Systems*, McGraw Hill, 1993.
33. Saunders, L., *The IBM VHDL design system*, Proc. 24th Design Automation Conference, 1987.
34. Schoen, J. M., *Performance and fault modeling with VHDL*, Englewood Cliffs, NJ: Prentice Hall, 1992.
35. Ward, P. C., and J. Armstrong, *Behavioral fault simulation in VHDL*, Proc. 27th Design Automation Conference, June 1990, pp. 587-593.

❑

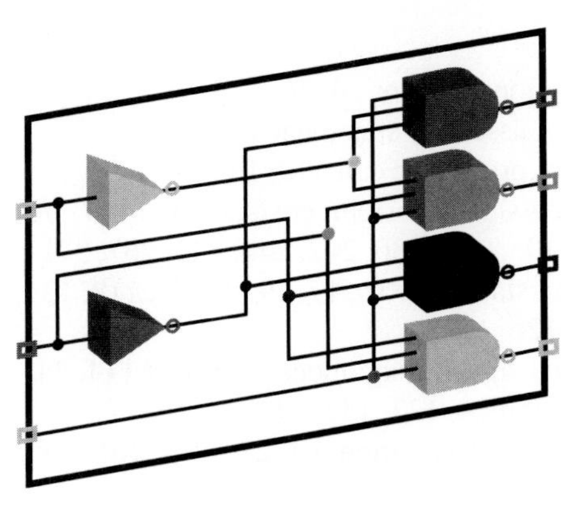

Index

T

U

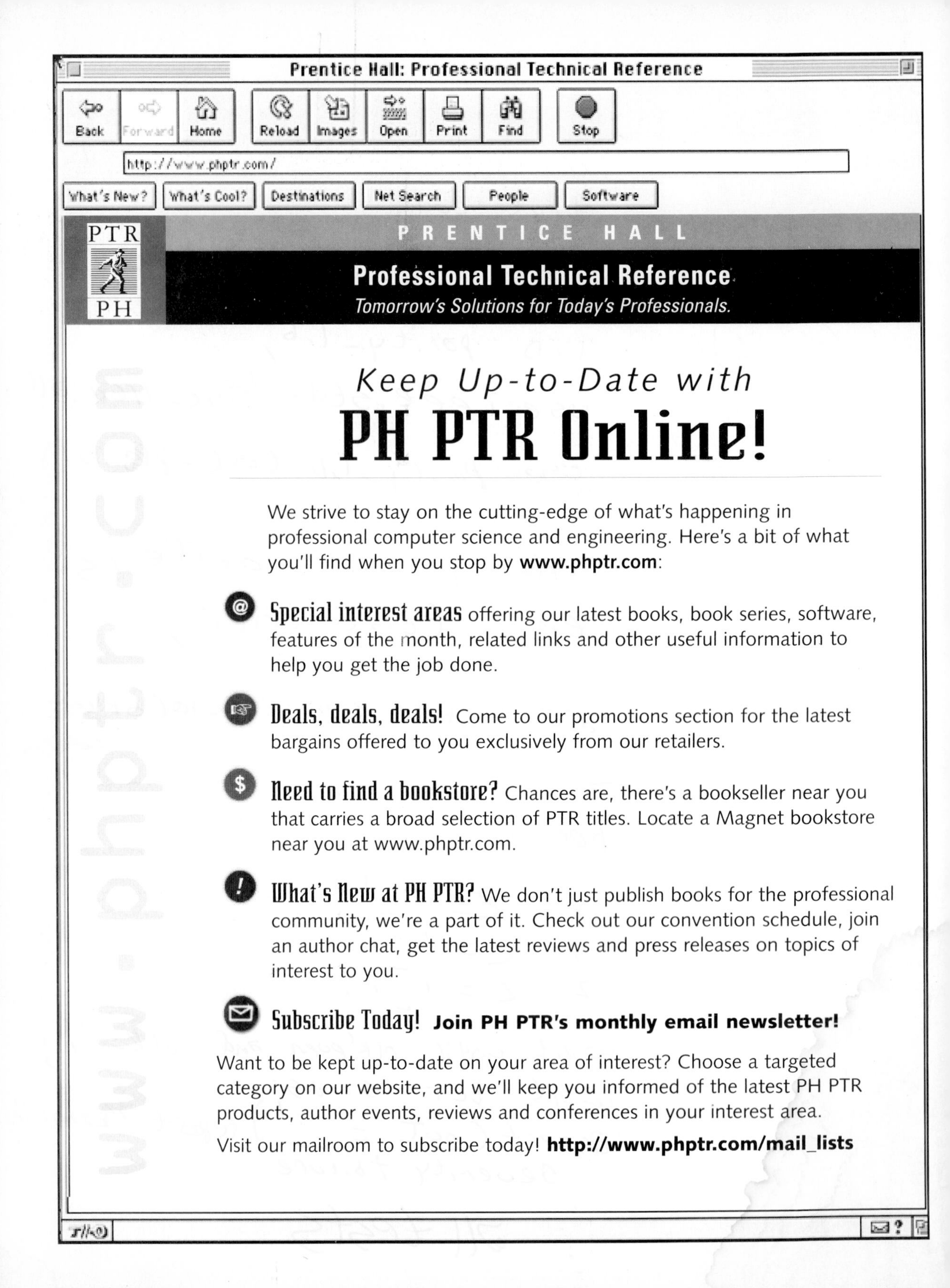

Prentice Hall: Professional Technical Reference
Back
Forward
Home
Reload
Images
Open
Print
Find
Stop
http://www.phptr.com/
What's New?
What's Cool?
Destinations
Net Search
People
Software
PTR
PH
PRENTICE HALL
Professional Technical Reference
Tomorrow's Solutions for Today's Professionals.
Keep Up-to-Date with
PH PTR Online!
We strive to stay on the cutting-edge of what's happening in professional computer science and engineering. Here's a bit of what you'll find when you stop by www.phptr.com:
Special interest areas offering our latest books, book series, software, features of the month, related links and other useful information to help you get the job done.
Deals, deals, deals! Come to our promotions section for the latest bargains offered to you exclusively from our retailers.
Need to find a bookstore? Chances are, there's a bookseller near you that carries a broad selection of PTR titles. Locate a Magnet bookstore near you at www.phptr.com.
What's New at PH PTR? We don't just publish books for the professional community, we're a part of it. Check out our convention schedule, join an author chat, get the latest reviews and press releases on topics of interest to you.
Subscribe Today! Join PH PTR's monthly email newsletter!
Want to be kept up-to-date on your area of interest? Choose a targeted category on our website, and we'll keep you informed of the latest PH PTR products, author events, reviews and conferences in your interest area.
Visit our mailroom to subscribe today! http://www.phptr.com/mail_lists
www.phptr.com

test code

```
lib ieee;
lib parity_lib;
use ieee.std_logic_1164.all;
use parity_lib.test_pkg.all

entity test_code is
end test_code;

architecture test_code_beh of test_
                                codeis

begin

 X  <= '_';
 Y  <= '_';
 Z  <= '_';
wait until clk'event and clk='1';
wait for 5 ns;
assert(out = _') report "Error"
   severity failure
 :
 :   all tests
 :
```

1725/mo

1
530 - Rent
400 - Food
120 - Gas
2 200 - Utes
1000 - Cars
700 - MC
* 1110 - LNA
160 - SP. Mall. 4

4221

400

2
1633
1633
700
700

4666

3% 15000
4500
2400

* 6900
1725

6900
4

29
28

100
4

20

Tags
Taxes
School
Backpack